# AN INTRODUCTION

TO

# PRACTICAL CHEMISTRY.

# AN INTRODUCTION

TO

# PRACTICAL CHEMISTRY,

## INCLUDING ANALYSIS.

BY

JOHN E. BOWMAN, F.C.S.,

PROFESSOR OF PRACTICAL CHEMISTRY IN KING'S COLLEGE, LONDON; AUTHOR OF "A HANDBOOK OF MEDICAL CHEMISTRY," ETC.

SECOND AMERICAN,

FROM THE SECOND AND REVISED LONDON EDITION.

PHILADELPHIA:

BLANCHARD AND LEA.

1856.

"Facts are the materials of Science; but all Facts involve Ideas. Since, in observing Facts, we cannot exclude Ideas, we must, for the purposes of Science, take care that the Ideas are clear, and rigorously applied."—WHEWELL, *Philosophy of the Inductive Sciences*, vol. i, p. xxxvii.

C. SHERMAN & SON, PRINTERS.
19 St. James Street.

# PREFACE

## TO THE SECOND EDITION.

---

In the present Edition I have endeavored, as far as possible, to make this little work more fully adapted to the requirements of the general student; and I believe it now leaves little to be desired as an Introductory Text-book of Practical Chemistry. I have also, without adding materially to its bulk, embodied such changes and corrections as have been rendered necessary by the progress of the science since the first edition was published.

J. E. Bowman.

King's College, London, February, 1854.

# PREFACE

## TO THE FIRST EDITION.

---

AMONG the many recent and valuable works on Chemistry, I am not aware of one having for its special object to explain, and render simple to the beginner, the various processes employed in analysis, or which have been devised for the illustration of the principles of the Science. Most of them contain much that is superfluous for the general student, who has but a limited time to devote to the subject; while they are wanting in those explanatory details, without which he must often fail to understand the rationale of the operations through which he is conducted.

It is with a wish to supply this deficiency, and at the same time to furnish a text-book for my own classes, that the present little work has been written; and as it is intended for the use of those who have made but little progress in the Science, my endeavor has been, through out, to make everything as simple and intelligible as possible. The employment of complicated or expensive apparatus has been almost wholly avoided.

The outline of most of the First Part was arranged some years ago by my friend Professor Miller (at that time Demonstrator of Chemistry in King's College), for the use and direction of the class of Chemical Manipulation, then first established to supply a growing demand, and to meet the requirements of the University of London, and some of the other examining Boards of the Metropolis. In the compilation of the Second and Third Parts, I have been much indebted to the excellent works of Rose, Fresenius, Parnell, and others; I must also here thank my colleague, whose name I have already mentioned, for many valuable suggestions, and for his kindness in revising the proof sheets, without which assistance many errors would have crept in, and rendered the book less worthy of the student's confidence.

JOHN E. BOWMAN.

KING'S COLLEGE, LONDON, September, 1848.

# CONTENTS.

## PART I.

### CHAPTER I.



### CHAPTER II.



### CHAPTER III.

CHAPTER IV.



CHAPTER V.



CHAPTER VI.



CHAPTER VII.



CHAPTER VIII.



---

# PART II.

## THE ACTION OF REAGENTS ON BASES AND ACIDS.

---

CHAPTER I.

## CHAPTER II.



## CHAPTER III.



## CHAPTER IV.



## CHAPTER V.

## CHAPTER VI.



## CHAPTER VII.



---

# PART III.

## QUALITATIVE ANALYSIS OF SUBSTANCES, THE COMPOSITION OF WHICH IS UNKNOWN.

---

## CHAPTER I.

## CHAPTER II.



## CHAPTER III.



## CHAPTER IV.



## CHAPTER V.



## CHAPTER VI.



## CHAPTER VII.

## CHAPTER VIII.



---

# PART IV.

## QUANTITATIVE ANALYSIS.



## CHAPTER I.



## CHAPTER II.

# PART V.

## CHAPTER I.

## APPENDIX.

# PRACTICAL CHEMISTRY.

## PART I.

### CHAPTER I.

INTRODUCTORY—SYMBOLS—EQUATIONS—GENERAL RULES.

1. So essentially is chemistry an experimental science, and so almost exclusively is it built up of facts which have been elucidated by experiment, that without experimental illustrations it would be quite impossible to teach or to study it with any great amount of success. It is not enough, however, for the student to see experiments performed by others; he must, if he would master even the general principles of chemistry, learn to make experiments himself; and he will, probably, be surprised how much more easily he will retain in his recollection those phenomena (as well as the principles they illustrate) which his own hands have been the means of producing. This is especially the case when he is enabled, while operating in the laboratory, to learn and study the theory of the changes which take place under his direction.

2. With the view of enabling the beginner to do this as much as possible, I have in the following pages explained, by means of chemical symbols and equations, nearly the whole of the changes and decompositions which take place in the experiments described. The symbols which I have made use of are those now almost universally adopted by chemists; and it will be seen by the following Table, that they consist, for the most part, of the first letter or two letters of the Latin names of the elements which they express.

*Table of Elementary Substances (arranged alphabetically), showing their symbols, atomic weights, and the composition of some of their compounds.*

| Name. | Symbol. | Atomic Weight. | Compounds. |
|---|---|---|---|
| Aluminum, . . | Al | 14 | $Al_2O_3$ Alumina.<br>$Al_2Cl_3$ Chloride of aluminum.<br>$Al_2O_3,3SO_3$ Sulphate of alumina. |
| Antimony, . . | Sb | 129 | $SbO_3$ Oxide of antimony.<br>$SbO_4$ Antimonious acid.<br>$SbO_5$ Antimonic acid. |
| Arsenic, . . . | As | 75 | $AsO_3$ Arsenious acid.<br>$AsO_5$ Arsenic acid. |
| Barium, . . . | Ba | 69 | BaO Baryta.<br>BaCl Chloride of Barium. |
| Bismuth, . . . | Bi | 107 | $Bi_2O_3$ Oxide of bismuth.<br>$Bi_2O_3,3NO_5$ Nitrate of bismuth.<br>$Bi_2Cl_3$ Chloride of bismuth. |
| Boron, . . . . | B | 11 | $BO_3$ Boracic acid. |
| Bromine, . . . | *Br* | 78 | $BrO_5$ Bromic acid.<br>HBr Hydrobromic acid. |
| Cadmium, . . | Cd | 56 | CdO Oxide of cadmium.<br>CdS Sulphide of cadmium. |
| Calcium, . . . | Ca | 20 | CaO Lime.<br>CaCl Chloride of calcium. |
| Carbon, . . . | C | 6 | CO Carbonic oxide.<br>$CO_2$ Carbonic acid.<br>$CS_2$ Sulphide of carbon. |
| Cerium, . . . | Ce | 46 | CeO Oxide of cerium.<br>$Ce_2O_3$ Sesquioxide of cerium. |
| Chlorine, . . . | Cl | 36 | $ClO_5$ Chloric acid.<br>$ClO_7$ Perchloric acid.<br>HCl Hydrochloric acid. |
| Chromium, . . | Cr | 28 | $CrO_3$ Chromic acid.<br>$Cr_2O_3$ Oxide of chromium.<br>$Cr_2O_3,3SO_3$ Sulphate of chromium. |
| Cobalt, . . . | Co | 30 | CoO Oxide of cobalt.<br>$Co_2O_3$ Sesquioxide of cobalt. |
| Copper, . . . (Cuprum.) | Cu | 32 | $Cu_2O$ Suboxide of copper.<br>CuO Black oxide of copper.<br>$CuO,SO_3$ Sulphate of copper. |

| Name. | Symbol. | Atomic Weight. | Compounds. |
|---|---|---|---|
| Didymium, . . | D | ? | ? |
| Fluorine, . . . | F | 18 | $HF$ Hydrofluoric acid.<br>$BF_3$ Fluoboric acid. |
| Glucinum, . . | G | 7 (?) | $GO_6$ Glucina.<br>$GCl_6$ Chloride of glucinum. |
| Gold, . . . . (Aurum.) | Au | 200 | $AuO$ Oxide of gold.<br>$AuO_3$ Teroxide of gold.<br>$AuCl_3$ Terchloride of gold. |
| Hydrogen, . . | H | 1 | *HO* (or *Aq*) Water.<br>$HO_2$ Binoxide of hydrogen. |
| Iodine, . . . | I | 126 | $IO_5$ Iodic acid.<br>$HI$ Hydriodic acid. |
| Iridium, . . . | Ir | 99 | $IrO$ Protoxide of iridium.<br>$Ir_2O_3$ Sesquioxide of iron. |
| Iron, . . . . (Ferrum.) | Fe | 28 | $FeO$ Protoxide of iron.<br>$Fe_2O_3$ Sesquioxide of iron. |
| Lanthanum, . | Ln | 48 | $LnO$ Oxide of lanthanum. |
| Lead, . . . . (Plumbum.) | Pb | 104 | $PbO$ Protoxide of lead.<br>$Pb_3O_4$ Red oxide of lead.<br>$PbCl$ Chloride of lead. |
| Lithium, . . . | Li | 7 | $LiO$ Lithia.<br>$LiCl$ Chloride of lithium. |
| Magnesium, . . | Mg | 12 | $MgO$ Magnesia.<br>$MgCl$ Chloride of magnesium. |
| Manganese, . . | Mn | 28 | $MnO$ Protoxide of manganese.<br>$MnO_2$ Binoxide or black oxide.<br>$MnO_3$ Manganic acid.<br>$Mn_2O_7$ Permanganic acid. |
| Mercury, . . . | *Hg* | 202 | $HgO$ Protoxide of mercury.<br>$HgO_2$ Red oxide of mercury.<br>$HgCl$ Protochloride of mercury.<br>$HgCl_2$ Perchloride of mercury. |
| Molybdenum, . | Mo | 48 | $MoO_3$ Molybdic acid. |
| Nickel, . . . | Ni | 30 | $NiO$ Oxide of nickel.<br>$Ni_2O_3$ Sesquioxide of nickel. |
| Nitrogen, . . . | N | 14 | $NO_5$ Nitric acid.<br>$NO_2$ Binoxide of nitrogen.<br>$NH_3$ Ammonia. |

| Name. | Symbol. | Atomic Weight. | Compounds. |
|---|---|---|---|
| Osmium, . . . | Os | 99 | $OsO_4$ Osmic acid.<br>$OsO_2$ Binoxide of osmium. |
| Oxygen, . . . | O | 8 | |
| Palladium, . . | Pd | 54 | $PdO$ Protoxide of palladium.<br>$PdO_2$ Peroxide of palladium. |
| Phosphorus, . . | P | 32 | $PO_5$ Phosphoric acid.<br>$PO_3$ Phosphorous acid.<br>$PH_3$ Phosphuretted hydrogen. |
| Platinum, . . | Pt | 99 | $PtO$ Protoxide of platinum.<br>$PtO_2$ Binoxide of platinum. |
| Potassium, . . (Kalium.) | K | 40 | $KO$ Potash.<br>$KCl$ Chloride of potassium. |
| Rhodium, . . | R | 52 | $RO$ Protoxide of rhodium.<br>$R_2O_3$ Sesquioxide of rhodium. |
| Ruthenium, . . | Ru | 52 | $Ru_2O_3$ Sesquioxide of ruthenium. |
| Selenium, . . | Se | 40 | $SeO_3$ Selenic acid.<br>$HSe$ Hydroselenic acid. |
| Silicon, . . . | Si | 22 | $SiO_3$ Silicic acid. |
| Silver, . . . . (Argentum.) | Ag | 108 | $AgO$ Oxide of silver.<br>$AgCl$ Chloride of silver. |
| Sodium, . . . (Natronium.) | Na | 24 | $NaO$ Soda.<br>$NaCl$ Chloride of sodium. |
| Strontium, . . | Sr | 44 | $SrO$ Strontia.<br>$SrCl$ Chloride of strontium. |
| Sulphur, . . . | S | 16 | $SO_3$ Sulphuric acid.<br>$HS$ Hydrosulphuric acid. |
| Tantalium, . . (or Columbium.) | Ta | 185 | $TaO_2$ Oxide of tantalium.<br>$TeO_6$ Tantalic acid. |
| Tellurium, . . | Te | 64 | $TeO_3$ Telluric acid.<br>$HT$ Hydrotelluric acid. |
| Thorium, . . . | Th | 60 | $ThO$ Oxide of thorium.<br>$ThCl$ Chloride of thorium. |
| Tin, . . . . (Stannum.) | Sn | 59 | $SnO$ Protoxide of tin.<br>$SnO_2$ Peroxide of tin. |
| Titanium, . . | Ti | 24 | $TiO$ Titanic acid.<br>$TiCl_2$ Bichloride of titanium. |
| Tungsten, . . (Wolfram.) | W | 96 | $WO_3$ Tungstic acid. |

| Name. | Symbol. | Atomic Weight. | Compounds. |
|---|---|---|---|
| Uranium, . . | U | 60 | $UO$ Protoxide of uranium.<br>$U_2O_3$ Sesquioxide of uranium. |
| Vanadium, . . | V | 69 | $VO_3$ Vanadic acid. |
| Yttrium, . . . | Y | 32 | $YO$ Yttria.<br>$YCl$ Chloride of yttrium. |
| Zinc, . . . . | Zn | 32 | $ZnO$ Oxide of zinc.<br>$ZnCl$ Chloride of zinc. |
| Zirconium, . . | Zr | 34 | $Zr_2O_3$ Zirconia.<br>$ZrCl_3$ Chloride of zirconium. |

3. Each of these symbols expresses one equivalent or atom of the substance which it represents. Thus, H stands for one atom or equivalent of hydrogen; Cu for an equivalent of copper; *Hg* for one of mercury.

When a small figure is placed to the right of a symbol, rather below the line, it means that there is that number of equivalents of the substance present. Thus, $Pb_2$ means two equivalents of lead; $O_5$ five equivalents of oxygen; $H_{10}$ ten equivalents of hydrogen.

Two or more symbols placed together, signify that the elements which they represent are chemically united in the closest manner. Thus *HO* stands for water, which is a compound of one equivalent of hydrogen, and one of oxygen; $SO_3$ represents anhydrous sulphuric acid, composed of one equivalent of sulphur and three of oxygen; $C_{12}H_{10}O_{10}$ represents starch, which consists of 12 equivalents of carbon, 10 of hydrogen, and 10 of oxygen, chemically combined together.

When symbols are separated by a comma, they represent compounds which are held together by a force less strong than that which unites elements that have no such mark interposed. Thus $KO,SO_3$ means sulphate of potash, composed of potash and sulphuric acid. The constituents of sulphate of potash, therefore, are both compounds, and the affinity which unites the potassium with the oxygen, and the sulphur with the three equivalents of the same element, is supposed to be stronger than that which unites the acid with the base, since it is easier to break it up into potash and sulphuric acid, than into potassium, oxygen, and sulphur.

When the sign + is interposed, it indicates that the substances between which it is placed are united in a manner still less intimate. Thus, in crystallized carbonate of soda ($NaO,CO_2+10\ Aq$), we have sodium and oxygen in the soda, and carbon and oxygen in the carbonic acid, combined in the closest and strongest manner; the soda and carbonic acid thus formed are separated by a comma, showing that they are held together by what we may here call the second degree of affinity; while the 10 equivalents of water of crystallization, separated by the sign +, are held by a much weaker force, so feeble indeed that a very moderate heat is sufficient to expel them.

The sign + is used also to separate the symbols of substances which are entirely disunited, as when we wish to express a mixture of carbonate of lime and hydrochloric acid, we put it thus, $CaO,CO_2+HCl$.

A large figure placed immediately before a symbol, multiplies all the symbols as far as the next comma or + sign. Thus, in the common phosphate of soda ($2NaO,HO,PO_5$) there are two equivalents of soda, one of water and one of phosphoric acid, combined together. If a large figure were placed before the whole formula enclosed in brackets, thus, $5(2NaO,HO,PO_5)$ it would represent 5 equivalents of the entire salt.

4. It is really wonderful how much these little symbols are capable of expressing, and how often and completely they assist in simplifying and rendering intelligible even the most complicated chemical changes; for besides the information they convey relative to the composition of the substances which they express, they can be so combined in the form of equations, as to show in the most perfect manner, the various compounds which result during chemical decompositions. For this purpose, the symbols of the substances employed are placed together so as to form one side of the equation; on the other side are placed those of the substances which are produced during the decomposition; and as no atom of matter is lost during these transformations,

it necessarily follows that the value of both sides of the equation must be equal. For example, the decomposition of carbonate of lime by hydrochloric acid may be thus represented:—

$$CaO,CO_2+HCl=CaCl+HO+CO_2.$$

Here we place the symbols of carbonate of lime and hydrochloric acid on one side, and on the other those of chloride of calcium, water, and carbonic acid, which are produced during the decomposition; and it will be observed that on each side there are exactly the same number of equivalents, viz. 1 of calcium, 3 of oxygen, 1 of carbon, 1 of hydrogen, and 1 of chlorine.

5. I have ventured to introduce a slight modification of the usual mode of printing the symbols, which will enable the student to see at a glance whether the substances expressed are in the solid, liquid, or gaseous form.

Those in the solid state are printed in strong Roman type, as Pb, lead. Liquids, or substances in solution, are printed in strong italics, as ***HO***, water; and gases or vapors are represented by fine hair letters, as H hydrogen, HO, steam.

Thus in the above equation, liquid hydrochloric acid (***HCl***) is poured on solid carbonate of lime ($CaO,CO_2$); chloride of calcium (***CaCl***) is formed, which remains in solution, together with carbonic acid ($CO_2$) which passes off in the gaseous form.

6. It is very important that the student should at once begin to make careful notes of all the experiments he engages in. He should endeavor to do this in as concise and methodical a manner as possible, and he will find it very advantageous to make use of symbols in describing the substances he employs, and the changes which they undergo: he will thus be able to record much in a small space, and at the same time he will be making himself familiar with the composition of the substances with which he is experimenting.

7. When, as is often the case, especially in analytical experiments, there are several solutions and precipitates either filtering, digesting, or waiting till the operator has leisure to attend to them, it is necessary to mark

them in some way, to prevent confusion. This is easily done with small pieces of gummed paper, on which a letter or number may be written, corresponding with a similar reference mark in the note-book.

8. The student will soon learn by experience that he cannot be too methodical in his operations, or too careful in cultivating habits of neatness and cleanliness. The presence of a little saline or other impurity in a glass, owing to careless washing, or a little extraneous matter having been allowed to find its way into a bottle or test tube, may retard or spoil the result of whole days of labor.

"Much as the chemist may soil his fingers during his experimental occupations, he will soon learn the great importance of cleanliness to the success of his experiments. The regular course of his operations causes many kinds of matter to pass in succession through his hands; and many of the substances, which by mixture have exhibited the phenomena they were competent to occasion, and so far answered the purpose of the experiment, then become mere useless dirt. Their dismissal and entire removal when thus circumstanced become necessary, that they may not contaminate other bodies; and are as imperatively required, as was the care previously bestowed to prevent their contamination from extraneous matter.

"It is this rapid change in the character and relation of the substances with which the chemist works, that makes a constant attention to cleanliness essentially necessary. The very bodies which at one moment are carefully retained in vessels that have previously been cleansed with the most scrupulous attention, become the next in the situation of so much dirt, from which the vessel must be cleansed as perfectly and carefully, before they can be fit for another experiment, as they were for the reception of the now rejected matter. The results of numerous experiments relative to testing bodies in solution by reagents, are in many cases dependent on the employing of clean vessels. For instance, a portion of water examined in glasses which have been carelessly washed, may occasion a slight precipitate with nitrate of silver or chloride of barium, and thus seem to contain a chloride or a sulphate (403, 429), when the cause of the precipitate may be nothing more than portions of salts adhering to the vessel.

"In the same manner the purity of an acid or a test, is not unfrequently affected by the state of the bottle containing it, or by the dirty condition of glass rods dipped into it, or of the funnels through which it has been poured or filtered, or of the vessels used in its transference; and sometimes it is contaminated by laying the stopper of the bottle containing it in a dirty place. Nor is it only that kind of dirt or impurity which gives an evident tinge to what it adheres to, that is to be avoided, but also numerous colorless substances, as salts, solutions, &c.; and in a word, anything which differs from the principal substance itself, and is at the same time liable to be dissolved or mixed with it.

"In consequence of these liabilities, and their interference with experiments, it should be established as a general rule in the laboratory, that no apparatus, nor any vessel (except such as may be destined to a particular use, and is as convenient when with a little previously adhering matter as if it were clean), be put away in a dirty state. All vessels or instruments when resorted to, should be found fit for the nicest experiment to which they are applicable. Glass rods or stirrers should be preserved in a clean place; glasses, on a clean shelf; and stoppers, when taken out of bottles, should be laid upon clean surfaces. These attentions and regulations will be found always useful, at times essential; and they are generally more requisite and influential in minute chemistry, than in large experiments."[1]

9. It is easy to clean even the dirtiest vessel, provided it has not been allowed to remain long with the impurities adhering to it; this, indeed, should never be permitted, and is readily avoided by making it a rule never to leave work for the day until the whole of the soiled apparatus has been thoroughly washed, and left to drain during the night, ready for wiping the next morning. For most purposes of cleaning, water will

[1] Faraday's Chemical Manipulation, p. 523.

be found sufficient, especially when the dirt it still moist; and when mere rinsing does not remove it, gentle friction with moist tow and coal ashes, will, in most cases, prove effectual. When the form of the vessel to be cleaned is such as will not allow the introduction of the hand (as flasks, test-tubes, &c.), a piece of stick or wire, having a little tow wrapped round the end, will be found very convenient. Glasses or basins that have been set aside to drain, should, before using, be wiped with a dry clean cloth, to remove any adhering particles of dust or moisture. Bottles or flasks, when required to be perfectly dry inside, may, after most of the water has been removed, be easily dried by warming them gently, and blowing air into them through a glass tube, either with the bellows or from the lungs; in this way the water is converted into vapor, which is quickly removed by the current of comparatively dry air.

When a glass or dish is greasy, it should be first wiped as clean as possible with tow or a dry cloth, then moistened with a little strong potash, and, lastly, well washed and rinsed with water. When the dirt to be removed is resinous also, or tarry, the application of strong potash or sulphuric acid will generally act upon it in such a way, that subsequent washing with water, together with gentle friction with coal ashes, will render it quite clean. It often happens, especially when a glass has been allowed to dry in a dirty state, that an insoluble crust is formed on the surface, which is very difficult of removal by mechanical means, but readily yields on the application of a few drops of hydrochloric or some other acid. An instance of this is afforded by solutions of lime, which, on exposure to the air, frequently deposit a crystalline sediment of carbonate of lime, which adheres strongly to the glass, but instantly dissolves on the addition of the acid.

10. When thrown upon his own resources, the student will often find it of the utmost value to be able to substitute, in default of more perfect apparatus, the common things used in domestic life, which are to be found in every house, such as glasses, plates, cups, saucepans, &c. When in addition to these he has at his command a blowpipe, a small piece of platinum foil and wire, a

flask or two, a funnel, and a little glass tubing of different sizes, he will, with the exercise of a little ingenuity and contrivance, be able to go through a very considerable course of experimental chemistry. He may rest assured that it is no disadvantage, but rather the contrary, to be thus compelled to devise and construct for himself rude and extemporaneous forms of apparatus; and if he should require encouragement to persevere in spite of the scantiness of his resources, he need only be reminded that the majority of those whose names shine brightest in the annals of science, have laid the groundwork of their future eminence while placed under the most unfavorable circumstances. So it was with the great Davy;[1] so with Dalton, with Scheele, Faraday, Dumas, Liebig, and may others almost equally illustrious.

"Habits of correct and delicate manipulation very much facilitate experimental inquiries at all times. It is not in difficult researches only that it is desirable, but even in such common operations as testing for lime, or iron, or sulphuric acid, its advantages become manifest; for either time is shortened, or the apparatus considered as necessary is diminished, or effectual substitution is made for those that may be wanting, and thus the experiment becomes easy, where otherwise it would be considered impossible. Besides facilitating such inquiries, it also diminishes the expense both in materials and apparatus, and it produces beneficial habits in the mind, by exercising it both in invention and perception even in this subordinate part of its operations. 'Nothing,' as Dr. Johnson observes, 'is to be considered as a trifle, by which the mind is inured to caution, foresight, and circumspection. The same skill, and often the same degree of skill, is exerted in great and little things.'"[2]

[1] "His means, of course, were very limited; not more extensive than those with which Priestley and Scheele began their labors in the same fruitful field. His apparatus, I believe, consisted chiefly of phials, wine-glasses, and tea-cups, tobacco-pipes, and earthen crucibles; and his materials were chiefly the mineral acids and the alkalies, and some other articles which are in common use in medicine."—*Life of Sir H. Davy*, by John Davy, M.D., vol. i, p. 43.

[2] Faraday, op. cit. p. vi.

# CHAPTER II.

## PNEUMATIC CHEMISTRY.

11. The gas-holder used in the following experiments, which bears the name of its inventor, Mr. Pepys, consists of an upright hollow box *a*, usually made of

Fig. 1.

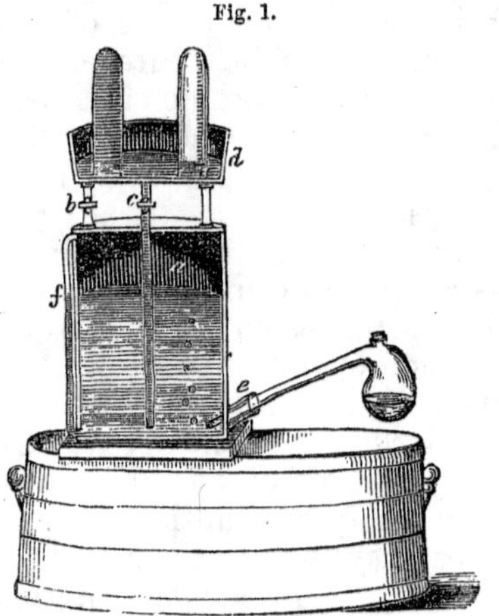


Section of Pepys' Gas-holder.

zinc or copper, connected by means of two tubes *b* and *c*, with a shallow pan *d* placed above it. The tubes are open at both ends, the longer one *c* reaching to within about half an inch of the bottom. The external glass tube *f* shows the height at which the liquid stands in the vessel. The lateral opening *e* is closed by a screw-cap before filling the vessel with water; the stopcoks *b* and *c* are then opened, and water poured into the upper pan, when it passes down the tube *c*, the air escaping up the other tube, until it is full, when no more bubbles of air will rise from the tube *b*.

When the gas-holder is filled with water, the stopcocks *b* and *c* are to be closed, and the screw-cap may be removed from *e* without danger of the water rushing out, since it is kept in by the pressure of the external atmosphere; but care must be taken not to remove the screw-cap while either of the stopcocks is open, as the water would rush out with great force. The beak of the retort may then be introduced, as shown, in the figure, when the gas will rise in bubbles through the water, which is gradually displaced, and flows out through the aperture *e*.[1]

## SECTION I.

### *Preparation of Hydrogen* (H).[2]

12. Weigh 300 grains of granulated zinc, and introduce the fragments carefully through the tubulure of the retort, sliding them, not dropping them in, to avoid the risk of breaking the bottom of the retort, which is usually of thin glass, and consequently seldom strong enough to bear a blow without injury.

Fig. 2.

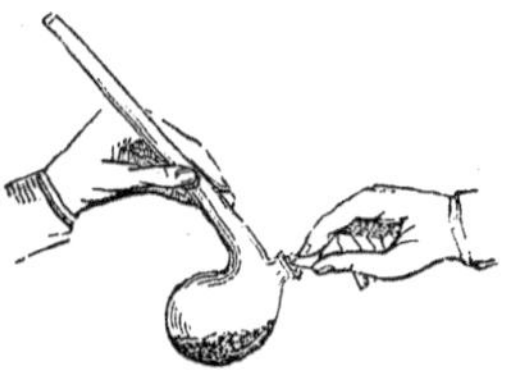

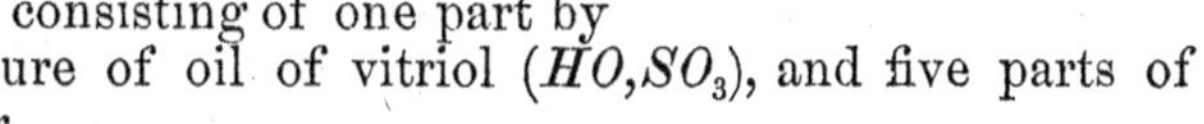

Pour upon the metal four fluid ounces of dilute sulphuric acid, consisting of one part by measure of oil of vitriol ($HO,SO_3$), and five parts of water.

Effervescence immediately commences, owing to the rapid evolution of the gas, the first portions of which, being mixed with the common air previously in the retort, may be collected separately in a small jar over the pneumatic trough, and afterwards rejected.

The nature of the decomposition that takes place may be seen in the following equation:

[1] In the absence of a gas-holder, the gases may be collected in jars over the pneumatic trough (see par. 18).

[2] The specific gravity of hydrogen is lower than that of any other form of ponderable matter, being only 0·069, that of common air being considered 1·000. 100 cubic inches weigh, at the ordinary temperature and pressure of the air, 2·138 grains, while the same quantity of common air weighs 31·00 grains. The atomic weight of hydrogen is 1, and its combining volume 1.

$$Zn + HO,SO_3 = ZnO,SO_3 + H.$$

The beak of the retort may now be inserted into the lateral aperture of the gas-holder, which should have been previously placed over the pneumatic trough so as to catch the water as it is displaced by the gas (Fig. 1).

When the effervescence subsides, and sufficient gas is collected, remove the gas-holder from the trough, and proceed with the following experiments.

Fig. 3.

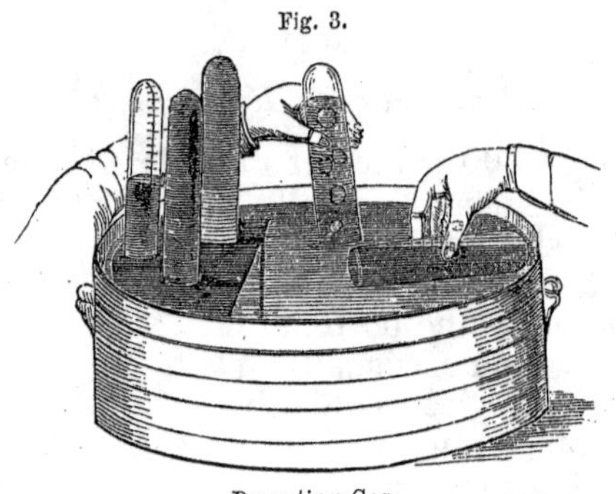

Decanting Gas.

13. Place an inverted jar filled with water, over the tube *b* (Fig. 1), and open both the stopcocks; the gas will immediately be forced upwards into the jar by the pressure of the water in the pan and in the tube *c*: close the bottom of the jar with a plate of glass, and remove it to the pneumatic trough. Decant a portion of this gas to a smaller jar (Fig. 3), and test its inflammability with a taper. Observe the deposit of dew in the inside of the jar after the combustion, which is the water formed by the combination of the hydrogen with oxygen. $H + O = HO.$

Fig. 4.

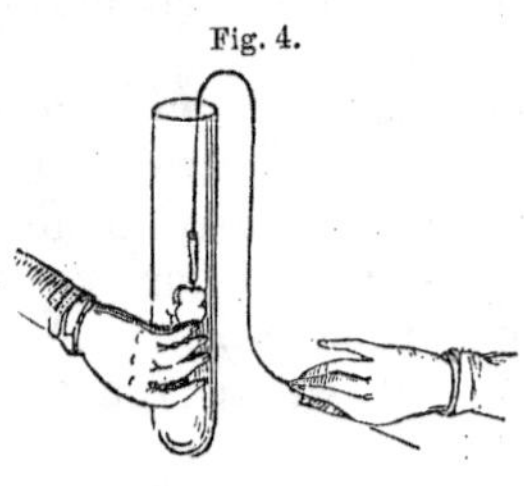

14. Fill a small jar with the gas, and having removed it from the gas-holder, let it stand for a few seconds with its open end upwards. If a lighted taper be now

applied, no combustion will ensue, as the hydrogen will have escaped upwards, on account of its very low specific gravity.

15. Repeat the last experiment, holding the jar with the open end *downward*. On applying a lighted taper, a slight explosion will take place, showing that the hydrogen had not escaped as before.

16. Transfer some of the gas from a large jar to a small one, and from this again to tubes, until it can be done without allowing any bubbles to escape. When the gas is to be decanted into a jar or tube which is much narrower, it may be first transferred into a lipped glass; or an inverted funnel may be used, as shown in Fig. 5.

Fig. 5.

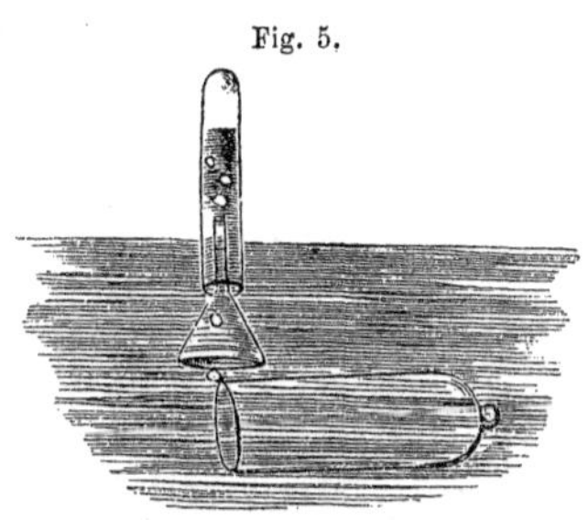

17. Transfer a little of the hydrogen in this way into a graduated tube, and mix it with varying but definite portions of common air; then ascertain by experiment what proportions detonate most loudly when a lighted taper is applied. The jars used for these experiments should be small and strong, to avoid risk of fracture by the force of the explosion.

## SECTION II.

### *Preparations of Carbonic Acid* ($CO_2$).[1]

18. As this gas is to a considerable extent soluble in water, it is better in its preparation not to use the gas-holder, on account of the large quanity of water it would then have to pass through, but to collect it at once in jars over the pneumatic trough. (Fig. 6.)

19. Put 300 grains of marble ($CaO,CO_2$) broken into fragments about the size of a pea, into a retort, observing the same precautions as were recommended in the

[1] The specific gravity of carbonic acid is 1·524 (air being 1·0), 100 cubic inches weighing 47·26 grains. Its atomic weight is 22; and its combining volume 1. At a temperature of 60° water dissolves about its own bulk of carbonic acid.

preparation of hydrogen (12). Measure out an ounce and a half of hydrochloric acid (*HCl*), dilute it with an equal quantity of water, and pour the mixture upon the

Fig. 6.

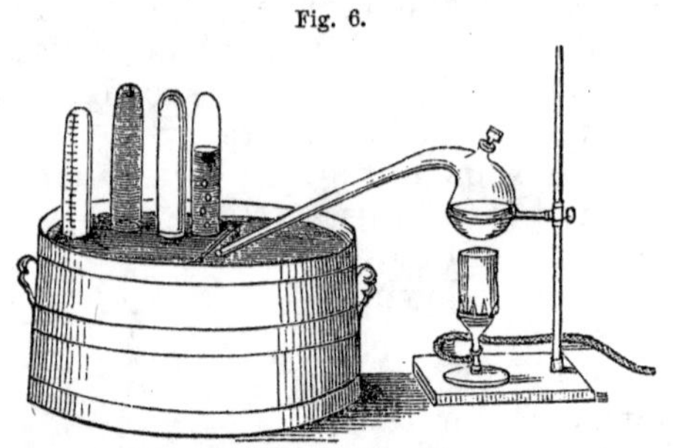

Pneumatic Trough.

marble. The gas is immediately given off, causing brisk effervescence, and it may be collected in jars placed on the shelf of the pneumatic trough, the first jarful being rejected as impure.

$$CaO,CO_2+HCl= CaCl+HO+CO_2.$$

20. Introduce a lighted taper into a small jar of the gas held with its open end upwards. It is instantly extinguished; and as the carbonic acid remains some time in the jar, on account of its high specific gravity, the taper may be extinguished repeatedly in the same jarful of gas.

21. Pour a little lime-water (*CaO*) into a test glass and thence into a jar filled with the gas, closing the mouth of the jar with a glass plate, and agitating the gas and liquid together. The lime-water almost immediately becomes milky, owing to the formation of carbonate of lime($CaO,CO_2$) which is insoluble in water.

Fig. 7.

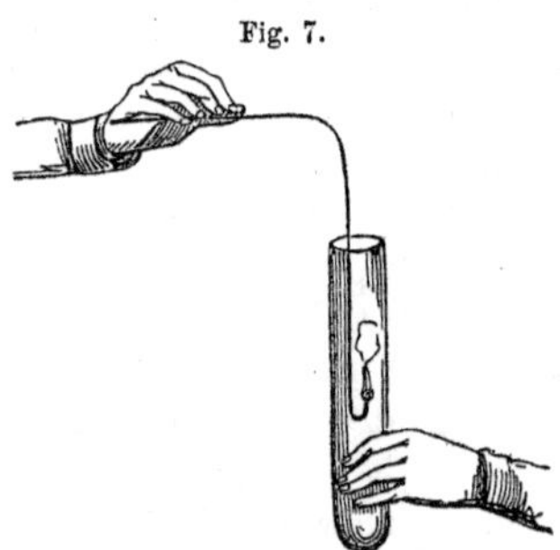

If a few drops of hydrochloric acid (*HCl*) be added, the carbonate of lime is decomposed, and the milkiness disappears, chloride of calcium being formed, which is soluble in water.

$$(CaO,CO_2+HCl=CaCl+HO+CO_2).$$

22. Having filled a jar with the gas, pour it like water into another jar somewhat smaller (Fig. 8); this is easily effected, owing to the high specific gravity of carbonic acid. Test its presence in both jars with lime-water (21), and by its power of extinguishing a taper (20).

Fig. 8.

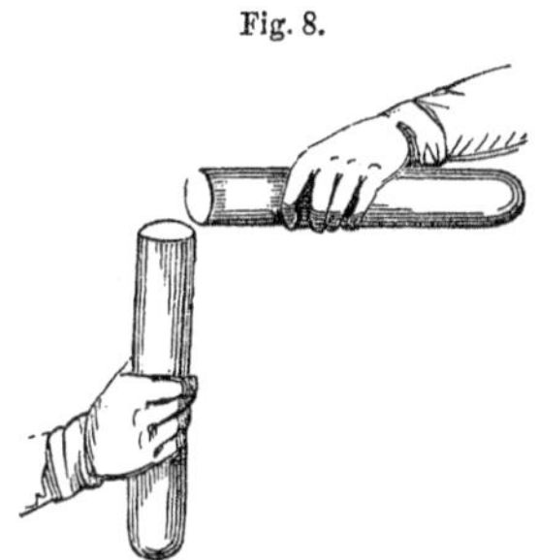

23. The high specific gravity of carbonic acid, and its power of extinguishing flame, may be strikingly shown by pouring it from a jar upon a lighted candle, which is instantly put out.

24. By means of a narrow tube open at both ends,

Fig. 9.

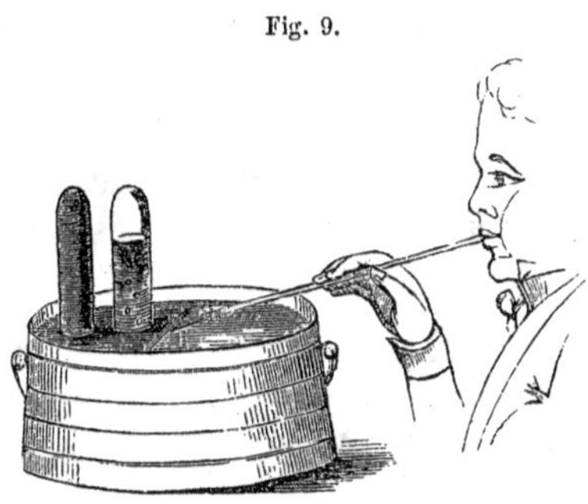

fill a jar over the pneumatic trough, with air from the lungs. Test it with a lighted taper, and observe that it causes an abundant precipitate in lime-water, owing to the presence of carbonic acid (21).

25. Invert a jar filled with common air over a lighted taper floating on the water of the pneumatic trough (Fig. 10); observe that it soon burns dim, and is shortly extinguished, the water at the same time slowly rising

Fig. 10.

in the jar. The absorption of air is here owing to the disappearance of the atmospheric oxygen, which combines with the hydrogen and carbon of the burning wax. Nearly one-fifth of the air is thus condensed, that being the proportion of oxygen contained in the atmosphere; the remaining four-fifths are nitrogen. When the combustion is over, invert the jar, and test the air contained in it with lime-water for carbonic acid (21).

## SECTION III.

### *Preparation of Binoxide of Nitrogen* ($NO_2$).[1]

26. Put 300 grains of copper turnings into a retort, and pour upon it an ounce and a half of strong nitric acid ($NO_5$) previously diluted with an equal quantity of water. Decomposition immediately commences, and the binoxide is formed by the action of the copper on a portion of the nitric acid, thus:—

$$3\,Cu+4NO_5=3(CuO,NO_5)+NO_2.$$

The gas which is first formed becomes orange, owing to its conversion into nitrous acid ($NO_4$) by combining with the atmospheric oxygen contained in the retort ($NO_2+2\,O=NO_4$).

27. Transfer a little to a jar, and test it with a taper; observe the orange fumes of nitrous acid which are instantly produced wherever the gas mixes with the air.

28. Measure a definite quantity of the gas in a graduated receiver, and transfer it to another jar over the pneumatic trough: then measure off an equal volume of atmospheric air, and add it by decantation, to the binoxide. When the orange fumes have disappeared, owing to the absorption of the nitrous acid by the water, transfer it again to the graduated jar, and ob-

[1] The specific gravity of binoxide of nitrogen is 1·039, 100 cubic inches weighing 32·22 grains. Its atomic weight is 30·0 and its combining volume 2.

serve the volume of the mixture, noticing accurately the difference between this and the sum of the original volumes employed before mixing. This experiment should be repeated three or four times, and if the results in each case agree pretty closely, take the average of the experiments, and the amount of condensation, divided by three, will give very nearly the quantity of oxygen contained in the atmospheric air employed. One equivalent of binoxide of nitrogen occupying two volumes, when combined with two equivalents of oxygen occupying one volume, forms one equivalent of nitrous acid ($NO_4$) which is absorbed by the water; consequently, one-third of the gas absorbed consists of atmospheric oxygen. If the experiment be carefully performed, the absorption will be found to be equal to about one-fifth of the volume of common air employed, that being the proportion of oxygen contained in it. (25).

Though the results obtained in this way are not very accurate, owing to the formation of other oxides of nitrogen, they are sufficiently so to allow of its occasional employment in determining the quantity of free oxygen in a gaseous mixture; and also when the whole of the uncombined oxygen has to be removed from a mixture containing it.

## SECTION IV.

### *Preparation of Olefiant Gas.* ($C_4H_4$).[1]

29. Pour into a retort six fluid drachms of alcohol ($C_4H_5O,HO$) and add to it in small portions an ounce and a half of strong sulphuric acid ($HO,SO_3$), gently agitating the mixture after each addition. Apply a moderate heat, and take care that the black froth which is formed towards the close of the operation does not boil over. Collect the gas in jars over the pneumatic trough, or in the gas-holder.

30. Examine a small jarful with a taper, and observe that, though the taper is extinguished, the gas burns with a bright white flame.

[1] The specific gravity of olefiant gas is 0·981, 100 cubic inches weighing 30·57 grains. Its atomic weight is 14, and its atomic volume 2.

31. When mixed with an equal volume of chlorine (Cl) the two gases combine, forming a heavy oily compound, called chloride of olefiant gas ($C_4H_4Cl_2$).

$$(C_4H_4)+2Cl=C_4H_4Cl_2.$$

The oil collects in drops on the sides of the jar and on the surface of the water, while the gases are gradually absorbed.

Olefiant gas derives its name from the circumstance of its forming this oily compound.

32. Mix together one volume of olefiant gas and two volumes of chlorine; close the jar with a glass valve, and quickly remove it from the pneumatic trough. Apply a light to the mixed gases, and observe the dense cloud of carbonaceous matter that is formed as the combustion gradually passes down the jar, hydrochloric acid being at the same time produced.

$$C_4H_4+4Cl=4HCl+4C.$$

## SECTION V.

### *Preparation of Carbonic Oxide* (CO).[1]

33. Carbonic oxide is prepared by the action of strong sulphuric acid ($HO,SO_3$) on oxalic acid ($HO,C_2O_3+2Aq$). When a mixture of the two acids is warmed, the oxalic acid is resolved into carbonic acid, carbonic oxide, and water, which latter unites with the sulphuric acid.

$$HO,C_2O_3+2Aq=CO_2+CO+3HO.$$

The carbonic oxide is purified from the carbonic acid by passing it through a solution of potash or milk of lime.

$$CO_2+CO+KO=KO,CO_2+CO.$$

34. Adapt a cork to a wide-mouthed bottle capable of holding half a pint of water, and fit to it two tubes (152), one of which, *a* (Fig. 11), should be about half an inch in diameter, straight and sufficiently long to reach nearly to the bottom; the other, *b*, should only just

[1] The specific gravity of carbonic oxide is 10·973, 100 cubic inches weighing 30·21 grains. Its atomic weight is 14·0, and its atomic volume 1.

pierce through the cork, and should be bent so as to deliver the gas, as shown in the figure: the diameter of this tube need not be more than about $\frac{1}{3}$ of an inch.

The beak of the retort may now be fitted with a cork, which should be bored to allow the bent tube *c* to pass through it; and care must be taken that this tube is sufficiently small to slide easily down the tube *a*, and long enough to reach to the bottom of the bottle.

Fig. 11.

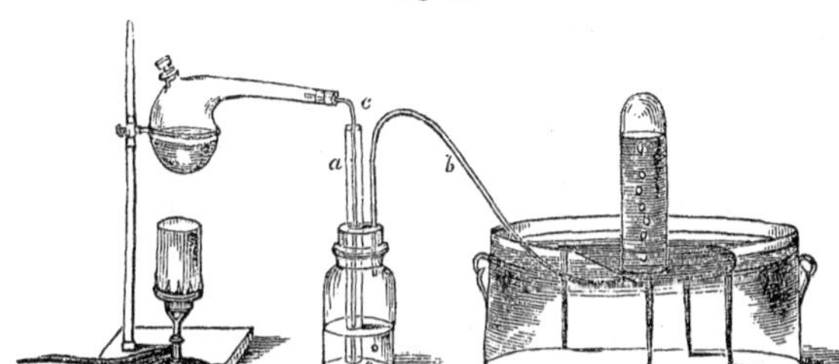


Preparation of Carbonic Oxide.

Four ounces of a tolerably strong solution of potash (*KO*) may now be introduced into the bottle.

35. Charge the retort with 180 grains of crystallized oxalic acid ($HO,C_2O_3+2Aq$) and two fluid ounces of strong sulphuric acid ($HO,SO_3$). On applying a gentle heat, the gas is given off, the first portions of which must be rejected as impure, and then two or three jars-ful may be collected over the pneumatic trough before the bottle containing the potash is connected with the retort. The gas thus obtained is a mixture of carbonic acid and carbonic oxide (33).

36. Having collected two or three jars full of the mixed gases for comparison, adapt the bent tube *c* to the mouth of the retort, and proceeded to purify the gas from carbonic acid, by passing it through the alkaline solution in the bottle. Pure carbonic oxide may then be collected.

37. Agitate a little lime-water with a jar full of the unpurified gas; the presence of carbonic acid is shown by the formation of carbonate of lime (21).

38. Repeat the experiment with a jar full of the purified gas. No precipitate ought now to appear.

39. Apply a lighted taper to a jar full of the impure gas, and observe the characteristic pale blue flame with which the carbonic oxide burns.

40. Do the same with a jar of the pure gas: the flame is brighter than when carbonic acid was present.

41. Pour a little lime-water into the jar used in the last experiment immediately after the combustion of the gas. The white precipitate which now appears, and which was not formed when the same gas was tested previous to the combustion, shows the result of that process to have been the formation of carbonic acid

$$CO+O=CO_2.$$

## SECTION VI.

### *Preparation of Oxygen* (O).[1]

42. Adapt a bent tube of the form shown in the figure, to a small hard glass flask, by means of a perforated cork.

Then weigh 100 grains of dried chlorate of potash

Fig. 12.

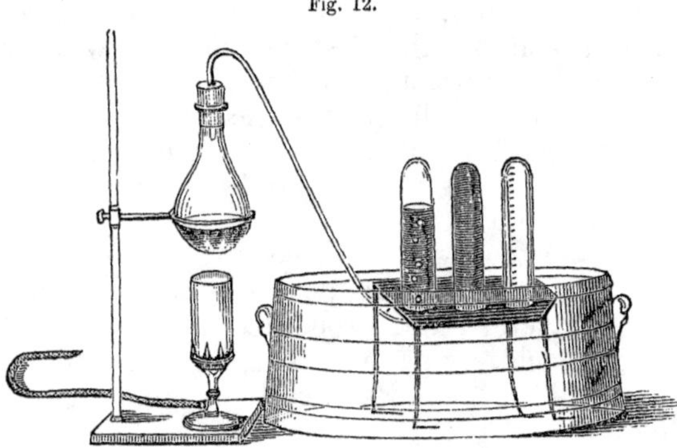

($KO,ClO_5$), mix it with 20 grains of black oxide of manganese ($MnO_2$), and place the mixture in the flask;

[1] The specific gravity of oxygen is 1·1057, 100 cubic inches weighing 34·29 grains. Its atomic weight is 8, and its combining volume ½.

adjust the tube so as to deliver the gas into the gas-holder, or under the shelf of the pneumatic trough (Fig. 12), and apply the heat of a lamp.

The chlorate of potash is thus decomposed, and gradually gives off the whole of its oxygen, which passes out through the tube, and may be collected either in the gas-holder or in jars, while chloride of potassium (KCl) remains in the flask, together with the oxide of manganese, which is not decomposed during the process.[1]

$$KO,ClO_5 = KCl + 6O.$$

The first portions of the gas should be rejected as impure, being mixed with the common air contained in the flask and tube.

43. The jars used for the following experiments should be open both at the top and bottom, the edges of both being ground smooth, so as to be closed air-tight with a glass valve *b* (Fig. 13).

Fig. 13.

*b*

*a*

44. Fill a jar with the gas, and introduce a glowing taper; it will instantly burst into flame, and burn with great brilliancy, until most of the oxygen is exhausted by combining with the carbon and hydrogen of the wax.

45. Introduce into another jar of the gas a small piece of ignited charcoal, attached to the end of a wire. It bursts into vivid combustion combining with the oxygen, and forming carbonic acid ($CO_2$), the presence of which may be proved by agitating a little lime-water in the jar (21).

46. Repeat the experiment with a small coil of thin iron wire, to which a little charcoal or amadou should be attached and ignited, for the purpose of heating the iron sufficiently to cause it to burn. The iron combines with the oxygen, forming the black oxide ($Fe_3O_4$), fused globules of which drop to the bottom, and should be received in water, as they are so intensely hot as to fuse into the glaze of a plate if allowed to fall upon it.

[1] The oxide of manganese is here used, because it is found that, when thus mixed, chlorate of potash gives off its oxygen with much greater facility and at a lower temperature than when heated alone.

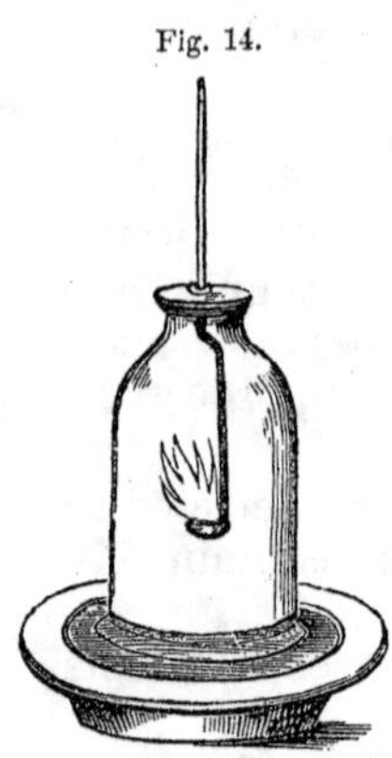
Fig. 14.

47. Place a fragment of sulphur about the size of a pea in the deflagrating spoon, set it on fire by holding it over a lamp, and introduce it into a jar of the gas; the sulphur burns with a brilliant blue flame, combining with the oxygen, and forming sulphurous acid ($SO_2$).

48. Mix together two volumes of hydrogen and one of oxygen, and with the mixture fill a small jar or tube, which for this experiment should be made of thick glass. On applying a light, the gases combine with a loud explosion, forming water. $H+O=HO$.

## SECTION VII.

*Preparation of Gases which are soluble in Water.*

49. Although in the preparation of many of the common gases it is most covenient to collect them over water, either in the gas-holder, or in jars placed in the pneumatic trough, still there are many cases in which this method is inapplicable, as when the gas is to any considerable extent soluble in water. It is usual in such cases, especially when great purity is necessary, to collect them in tubes or jars over mercury, which is not acted upon by the majority of the gases. For common purposes, however, some of them may be collected by the displacement of common air in dry bottles, and the more the gas differs in density from atmospheric air, the more is this method applicable.

Hydrochloric acid gas and ammonia may be taken as examples of the process.

*Preparation of Hydrochloric acid Gas* ($HCl$).[1]

50. This gas is easily obtained by the action of sulphuric acid on common salt.

[1] The specific gravity of hydrochloric acid gas is 1·269, 100 cubic inches weighing 39·37 grains. Its atomic weight is 37·0, and its atomic volume 1.

To the beak of a retort, a bent tube of the form represented in Fig. 15, is adapted by means of a perforated cork; a loose roll of filtering paper is introduced into the neck, to retain any moisture that may distil over;

Fig. 15.

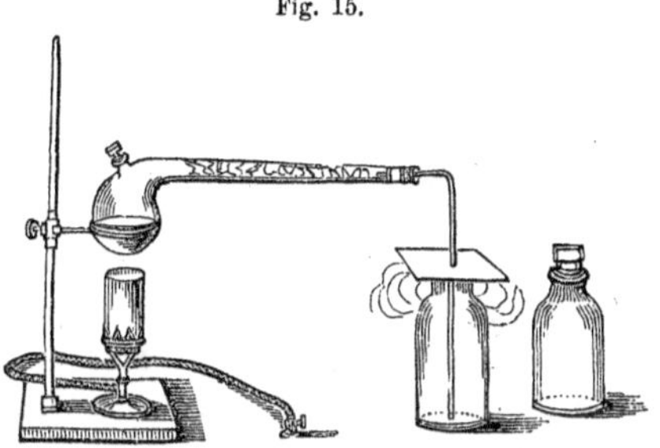

Preparation of Hydrochloric Acid Gas.

and the retort is charged by introducing 300 grains of dry chloride of sodium (NaCl), and adding to it six fluid drachms of strong sulphuric acid ($HO,SO_3$). Immediate effervescence takes place, and the bent tube is passed into a dry bottle of about a pint capacity, which should be furnished with a greased stopper; while the bottle is filling, the mouth may be loosely closed with a piece of card or paper.

Observe the dense fumes which are formed wherever the gas mixes with the air, especially if the atmosphere is damp, owing to the combination of the gas with the aqueous vapor. The bottle may be considered full when the gas has been flowing over from the mouth of the bottle for two or three minutes; the tube should then be cautiously withdrawn, and the bottle tightly closed with the stopper. Three or four bottles may be similarly filled with the gas, a gentle heat being applied if necessary.

The decomposition may be thus represented:—

$$\mathrm{NaCl}+HO,SO_3=\mathrm{NaO,SO_3}+\mathrm{HCl}.$$

The sulphate of soda of course remains in the retort.

51. Ascertain the action of the gas on a lighted taper.

52. Remove the stopper from one of the bottles, in-

stantly close it again with a dry glass plate (a precaution which is on no account to be omitted, as the stopper might in that case become immovably fixed), and plunge it with the mouth downwards into the water of the pneumatic trough. If the bottle has been well filled, the water will, when the glass plate is removed, quickly raise and nearly fill it, while the unabsorbed residue shows the quantity of common air left in the bottle.[1]

This experiment must not be made without first removing the stopper, and substituting the glass plate; if it is attempted to take out the stopper while the bottle is under water, there is great danger of its becoming so firmly fixed, as to be almost incapable of removal, owing to the absorption of the gas by the water, and the formation of a partial vacuum.

53. Test a little of the acid solution obtained in the last experiment, in a tube, with litmus paper, and afterwards with a few drops of solution of nitrate of silver ($AgO,NO_5$). The white precipitate, which is chloride of silver ($AgCl$), will be found to be insoluble in nitric acid, but readily soluble in ammonia (429).

54. Reserve a bottle of the gas for an experiment (60) with ammonia.

## SECTION VIII.

### *Preparation of Ammoniacal Gas* ($NH_3$).[2]

55. This gas may be prepared in a similar manner to the last; but as it is specifically lighter than common air, the bottles in which it is collected must be kept, while filling, with the mouth downwards, the delivering tube passing upwards, to the top (Fig. 16); the neck of the retort should be furnished as before with a roll of filtering paper.

56. Reduce 300 grains of quick lime ($CaO$) to powder in a mortar, and slake it in a small basin with a drachm

[1] Water at common temperatures is capable of dissolving no less than 480 times its own volume of hydrochloric acid. The liquid hydrochloric or muriatic acid of commerce is a solution of the gas in water.

[2] The specific gravity of ammoniacal gas is 0·589, 100 cubic inches weighing 18·288 grains. Its atomic weight is 17, and its atomic volume 2.

and a half of water; then pound 400 grains of muriate of ammonia ($NH_4Cl$); mix the powders as quickly as

Fig. 16.

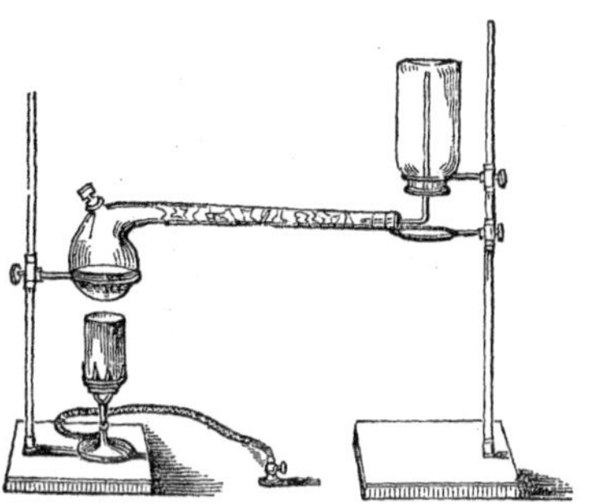

Preparation of Ammoniacal Gas.

possible, and without loss of time transfer the mixture to the retort. The following decomposition takes place:—

$$NH_4Cl + CaO = CaCl + \mathit{HO} + NH_3.$$

If the gas does not come over rapidly, a gentle heat may be applied. When three or four bottles have been filled, proceed with the following experiments:—

Fig. 17.

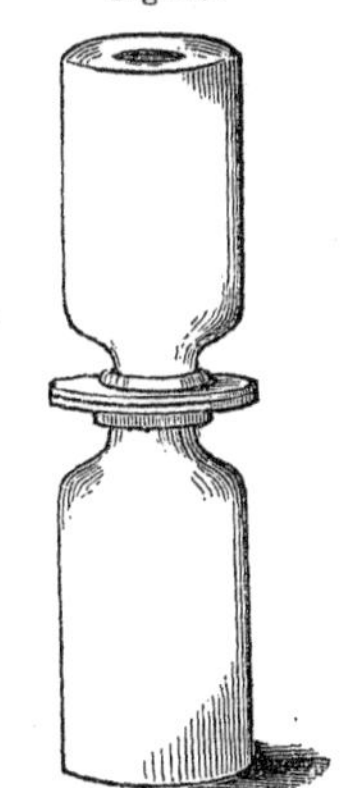

57. Observe the effect of the gas on a lighted taper: it extinguishes the flame, and at the same time shows a slight tendency to burn with a pale green flame.

58. Remove the stopper from one of the bottles, and close the mouth with a dry glass plate (read paragraph 52); then invert it, and having placed it under water, remove the glass plate and observe the rapid absorption. That which remains unabsorbed is atmospheric air.[1]

59. Test the liquid obtained in the last experiment (which is a weak solution of ammonia) with turmeric and red-

[1] Water at common temperatures is capable of absorbing nearly 700 times its volume of ammoniacal gas.

dened litmus-paper; the first is turned brown, the latter has the blue color restored.

60. Remove the stopper from a bottle of the gas, and also from the reserved bottle of hydrochloric acid (54), replacing them with dry glass plates. Then invert the latter over the bottle of ammonia (Fig. 17), and cautiously remove the glass plates so as to allow the gases to mix. Dense white fumes, consisting of muriate of ammonia ($NH_4Cl$) are immediately produced, which in a short time collect in flakes, and fall like snow on the sides and bottom of the vessels. In this combination of the hydrochloric acid with the ammonia, considerable heat is evolved.

---

## CHAPTER III.

### DISTILLATION.

#### SECTION I.

*Distillation of Water.*

61. ADAPT a cork to the neck of a quilled receiver, and bore a hole through it to fit the neck of the retort, which should pass through it for about two inches.

Fig. 18.

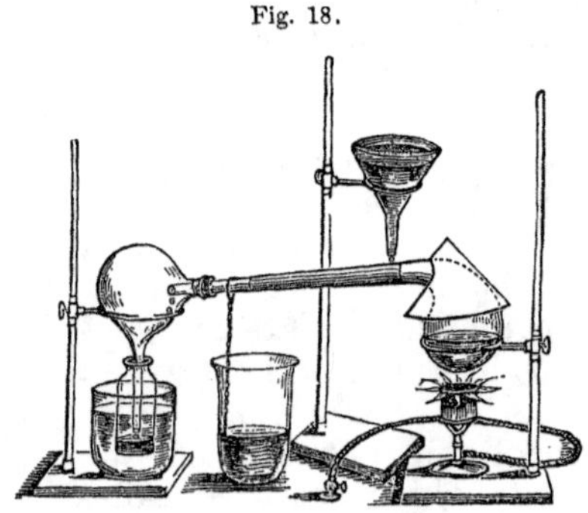

When this is done, the apparatus may be fitted up as shown in the figure (Fig. 18). The funnel which supplies water for cooling the neck of the retort, has its throat partially obstructed by a plug of tow, to regulate the flow of liquid; the neck of the retort is covered by a slip of bibulous paper of the form of the annexed sketch (Fig. 19), cut of such a width as almost completely to encircle the neck; and between the lower end of the paper and the quill receiver a thin fillet of tow is twisted tightly round the glass, to carry off the superfluous water, which drops into a basin placed underneath for its reception. The quill of the receiver passes into a small flask or bottle, which is kept immersed in water during the process, in order to keep it cool.

Fig. 19.

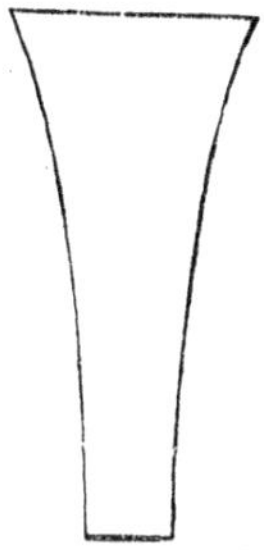

62. When the apparatus is thus arranged, the retort must be cautiously charged with common water till nearly half full, care being taken that none of it gets into the neck, as it would run down into the receiver and contaminate the distilled water, which should otherwise be pure. The upper part of the body of the retort being then covered with a conical cap of paper to prevent loss of heat by currents of air and radiation, the lamp may be applied, care being taken that the ebullition does not go on too violently, lest any of the impure water should splash or boil over into the neck of the retort. If, instead of boiling quietly and uniformly, the water in the retort "bumps," owing to the sudden disengagement of large bubbles of steam, a few fragments of broken glass or platinum wire may be placed in the retort to assist the formation of small bubbles from their surface. The first ounce of water that comes over should be rejected as impure, after which two or three ounces may be distilled for examination.

Fig. 20.

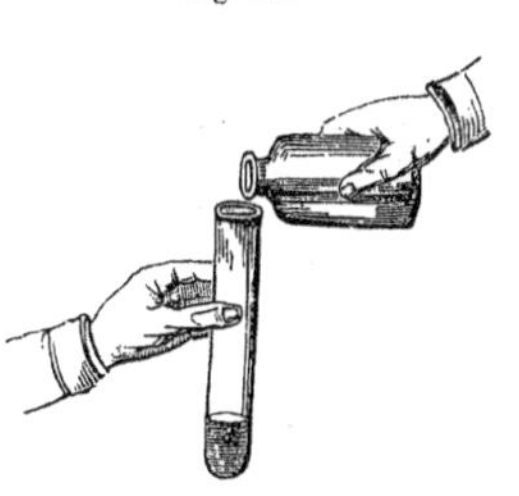

63. While the distillation is going on, another portion of the water operated on may be tested, with the view of dis-

covering some of the impurities present in it. Fill four test-tubes about one-third full of the undistilled water, and add to them respectively a few drops of the following reagents.

(*a.*) To the first add a solution of *chloride of barium* ($BaCl$); a white precipitate, insoluble in nitric acid,[1] indicates the presence of sulphates (403), most commonly sulphate of lime ($CaO,SO_3$).

(*b.*) To another portion add a solution of *nitrate of silver* ($AgO,NO_5$). If any chloride is present, usually chloride of sodium ($NaCl$), a white curdy precipitate of chloride of silver (AgCl) will be produced, insoluble in nitric acid, but readily soluble in ammonia (429). By exposure to the light this precipitate gradually becomes purple, especially when the water contains organic matter.

(*c.*) To the third tube add a little *lime-water* ($CaO$ in water): a white precipitate, soluble in nitric acid, shows that carbonic acid ($CO_2$) is present (420).

(*d.*) To the remaining tube *oxalate of ammonia* ($NH_4O,C_2O_3$) may be added, which will give a white precipitate if any lime is present (218).

64. Test the distilled water in the same way; if pure it will of course furnish no precipitate with any of the reagents.

65. Evaporate a few drops both of the distilled and undistilled water on platinum foil or a clean slip of glass: a considerable residue will probably be left by the latter, but no trace of solid matter ought to be observable where the other lay.

66. During ebullition, the water in the retort usually becomes turbid, owing to the formation of a white insoluble powder, which may be separated by filtration when the distillation is over.

To prepare a filter, take a small piece of white filtering or blotting paper, and fold it twice from side to side (Fig. 21); then round off with scissors the projecting corners, so that the paper may fall wholly within the

[1] In testing the solubility of a precipitate in any liquid, pour off a small portion into a separate tube for the experiment, reserving the rest for comparison.

funnel (Fig. 22). Moisten the paper placed in a funnel with distilled water, and then carefully pour in the li-

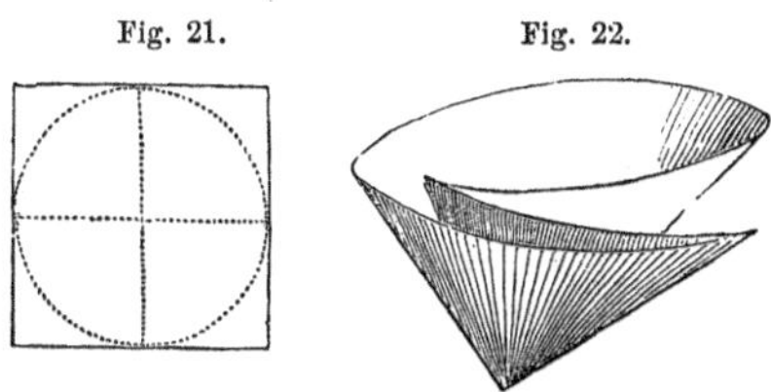

Fig. 21. Fig. 22.

quid to be filtered, using a glass rod to conduct it (Fig. 23), (636).

When most of the liquid has passed through, the white powder may be detached with a knife from the

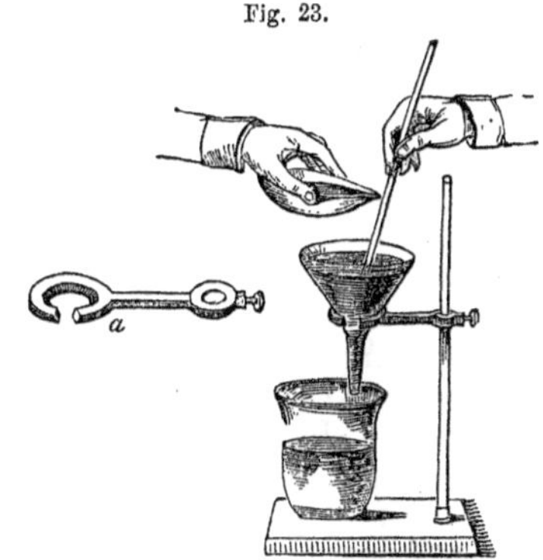


Fig. 23.

paper, and introduced into a test-tube; the clear solution being reserved for examination (68).

67. (*a.*) Add a few drops of dilute *nitric acid* to the powder in the tube, and observe that it dissolves with effervescence, indicating that it is a carbonate (419).

(*b.*) Supersaturate the solution thus obtained with ammonia, and add a little *oxalate of ammonia* ($NH_4O, C_2O_3$): a white precipitate shows the presence of lime (218). The powder is thus proved to be carbonate of lime ($CaO,CO_2$). This carbonate of lime had been held in solution by the excess of carbonic acid contained in the water; when the gas is expelled during ebullition, the carbonate is precipitated.

68. Test the solution filtered from the carbonate of lime in (66) with *chloride of barium*, *nitrate of silver*, *lime-water*, and *oxalate of ammonia*; and compare the results with those obtained in (63), when the water was examined in its natural state. As most of the lime has been separated as carbonate, we may expect to find less of it in solution than before, but more of the sulphates and chlorides, since they still remain dissolved in a more concentrated form.

## SECTION II.

### *Distillation of Liquid Hydrochloric Acid* (HCl in water).

69. Fit up the apparatus as in the ordinary process of distilling water (61), taking care that all the joints are perfectly tight: then remove the retort, and introduce through the tubulure 1000 grains[1] of dry chloride of sodium (NaCl) in coarse powder, taking care that none of the particles fall into the neck of the retort; then adjust the apparatus as before. Measure into the small flask or bottle which is to receive the distilled acid, 12 fluid drachms of water, and mark with a file or a strip of waxed paper, the height at which it stands; and having emptied it, measure into it seven drachms of distilled water. During the distillation care must be taken that the quill of the receiver dips under the surface of this water, which will assist in condensing the acid fumes, some of which might otherwise escape.

Into a small evaporating basin, pour seven drachms of water, and add gradually to it six drachms of strong sulphuric acid ($HO,SO_3$), stirring the mixture with a glass rod. When nearly cool, this dilute acid may be poured carefully into the retort through a small funnel, avoiding any splashing or soiling of the neck. A gentle heat may then be applied, which must be regulated ac-

[1] In this and many of the other experiments, small quantities are mentioned to suit the convenience of my class of Practical Chemistry, the lessons being only two hours long. When the products of the experiments are wanted for use, much larger quantities must frequently be employed.

cording to the rapidity with which the acid distils over, great care being taken that the mixture does not boil over into the neck of the retort (50).

The distillation may be continued until twelve drachms of acid have come over, which may be known by the mark previously made in the receiving flask.

70. The acid in the receiver may now be examined as to its purity. Pour a little into a test-tube, dilute it with about three times its bulk of water, and add a few drops of a solution of *chloride of barium;* if a white precipitate appears which is insoluble in the acid, it shows the presence of sulphuric acid as an impurity (403).

71. Evaporate a few drops of the acid on platinum foil or a clean slip of glass: no trace of the spot where it lay ought to remain. Any solid residue shows the presence of some saline impurity, caused probably by a little of the salt employed having got into the neck of the retort, and been washed down into the receiver.

## SECTION III.

### *Distillation of Liquid Ammonia* ($NH_3$ in water).

72. Prepare the apparatus as in the distillation of hydrochloric acid (69).

Pound 450 grains of quick-lime (CaO), introduce it into the retort through the tubulure, and pour gradually upon it two ounces of distilled water. Measure into the receiving flask or bottle fifteen drachms of water, and mark with a file or waxed paper the height at which it stands; empty it, and pour in two drachms of distilled water for the quill of the receiver to dip into during the distillation.

Weigh out 530 grains of muriate of ammonia ($NH_4Cl$), dissolve it in three ounces of water in a small evaporating basin, and pour the solution into the retort.

The distillation may now be commenced, carefully regulating the heat, and continuing it until the distilled liquid reaches up to the file mark in the receiver, when 15 drachms will have been collected.

$$CaO + \mathit{NH_4Cl} = NH_3HO + \mathit{CaCl}.$$

73. Pour a little of the ammoniacal solution thus prepared into a test-tube, and add to it a few drops of *chloride of barium:* if a precipitate appears, it is owing to the presence either of carbonic or sulphuric acid. To distinguish between them, add *nitric acid* in slight excess; if the precipitate thereupon dissolves, it is carbonic acid (421); if not, it is sulphuric (403).

74. Test another portion of the ammoniacal solution with a little *oxalate of ammonia;* if a white precipitate is formed, it is owing to the presence of lime as an impurity (218).

75. Supersaturate a little of the distilled liquid with nitric acid in a test-tube, and add a few drops of a solution of *nitrate of silver* ($AgO,NO_5$); a white precipitate indicates the presence of hydrochloric acid or a chloride. If a further portion of the ammoniacal solution be added, so as to render the liquid alkaline, the precipitate redissolves (429).

76. If no precipitate occur with any of these tests, evaporate a few drops of the ammoniacal solution on a slip of glass or platinum foil, and observe whether any trace of saline impurity is left.

## SECTION IV.

### *Distillation of Liquid Nitric Acid* ($NO_5$ in water).

77. Fit up the apparatus as in the distillation of hydrochloric acid (69). Introduce into the retort 1000 grains of nitrate of potash ($KO,NO_5$); pour upon it ten drachms of strong sulphuric acid ($HO,SO_3$) previously diluted with an equal bulk of water, and apply a gentle heat, observing the same precautions as were recommended in the former cases (61, 69).

$$KO,NO_5+2(HO,SO_3)=KO,SO_3,HO,SO_3+HO,NO_5.$$

78. While the distillation is going on, dissolve a few crystals of the nitrate of potash in distilled water, for the purpose of ascertaining its purity.

(*a.*) Test a little of the solution with *nitrate of silver* ($AgO,NO_5$); if any chloride is present, a white curdy precipitate appears, which is insoluble in nitric acid, but readily soluble in ammonia (429).

If the nitrate employed is contaminated with any chloride, the acid that distils over is sure to contain a little hydrochloric acid ($HCl$).

(*b.*) To another portion, add a solution of *chloride of barium* ($BaCl$); if any sulphates are present, a white precipitate is produced, which is insoluble in nitric acid (403).

79. Dissolve a small quantity of nitrate of potash in hot water, in an evaporating basin, adding the salt as long as it is taken up by the water on stirring; pour the hot solution into another basin, and observe the gradual formation of crystals as it cools. Remove some of these from the liquid, and dry them on filtering paper; then redissolve them in distilled water, and test the solution as before, with *nitrate of silver* and *chloride of barium*. The precipitates, if any, will be less dense than in the previous examination, showing that a partial purification has been effected.

Fig. 24.

80. The distilled nitric acid may now be tested for impurities, but before the test liquids are applied a portion should be diluted with four or five times its bulk of distilled water, since the chloride of barium is itself insoluble in strong nitric acid, and would consequently cause a precipitate, even though no sulphuric acid were present. A portion may then be tested for sulphates and chlorides with *chloride of barium* and *nitrate of silver* (403, 429).

81. If the distilled acid is found to contain sulphuric or hydrochloric acids, it may be purified from them by adding a solution of nitrate of silver as long as any precipitate is produced, and re-distilling; when those acids will remain behind in combination with the oxide of silver.

$$HCl+HO,SO_3+2(AgO,NO_5)=\mathrm{AgCl+AgO,SO_3+2(HO,NO_5)}.$$

# CHAPTER IV.

## GLASS-WORKING.

82. THE most convenient form of apparatus for working glass on the small scale, is the water blowpipe, which consists of an upright box, about fifteen inches high, of the form represented in Figure 25. It is usually made of zinc or copper, and is divided into two compartments by the plate *a*, which passes down to within about half an inch of the bottom, thus leaving a communication open between the two. The lower end of the tube *b* is closed by a valve opening outwards, to prevent the escape of air in that direction: the box should be filled about half full of water, and when used, air is blown through the tube *b*. The pressure thus occasioned in the compartment *c*, forces a portion of the water into the next division *d*, where it rises to a higher level than in *c*, and by its superior pressure forces a stream of air through the fine aperture at the extremity of the tube *e*, as long as it continues to stand at a higher level than in *c*. In this way a continuous jet is readily obtained, with much less fatigue to the operator than with the mouth blowpipe.

Fig. 25.

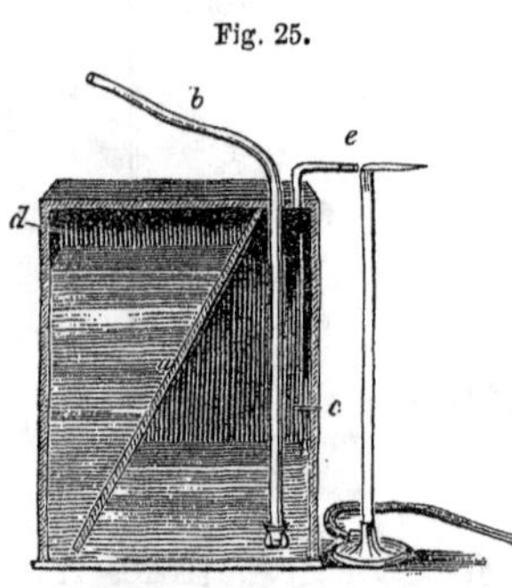

Water Blowpipe.

83. If the blowpipe flame be examined, it will be found to consist of two distinct parts, which may be called, for the sake of distinction, the inner *a*, and the outer flame *b* (Fig. 26). The blue point of the inner flame is evidently surrounded on all sides by the burning gas, no atmospheric oxygen being near it, so that any substance containing oxygen loosely combined,

placed in it, will be decomposed by the powerful deoxidizing affinities of the carbon and hydrogen of the combustible gases: on this account the inner flame is usually called the *deoxidizing* or *reducing* flame. The outer flame, on the contrary, is surrounded on all sides by the external air, so that here there is no excess of combustible or deoxidizing matter, but rather an excess of atmospheric oxygen; so that an oxidized substance may be placed at its extremity without danger of deoxidation, unless such decomposition is effected by the mere heat of the flame, independent of its chemical action; on the other hand, most substances, having an affinity for oxygen, placed within its influence, become oxidized at high temperatures, and hence it is usually called the *oxidizing* flame.

Fig. 26.

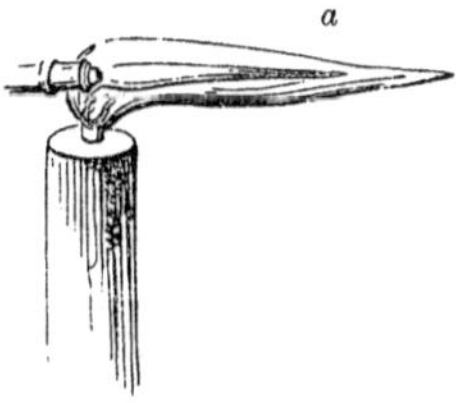

a

Blowpipe Flame.

84. The English flint glass, of which the tubes and rods commonly in use are made, contains in its composition a quantity of oxide of lead (PbO), which, when heated in contact with deoxidizing matter, is very easily decomposed. On this account it is necessary, in heating glass with the blowpipe, to take care that it does not approach the deoxidizing flame, but is kept at the extremity of the oxidizing flame, otherwise a black stain of metallic lead will be deposited on the surface of the glass. Slight stains of this description may generally be removed by holding the glass for a few seconds in the oxidizing flame; this converts the lead again into oxide, which dissolves in the glass.

85. *Make a few glass stirring rods, of lengths varying from five to eight inches.* To do this, a piece of solid rod, long enough to make two stirrers, should be held at a short distance from the extremity of the flame, and gradually brought towards it; a rotating motion being communicated to it by means of the finger and thumb, so that the part where the heat is applied may be uniformly heated all round (Fig 27). When the glass

begins to soften, it should be gently pulled with both hands, until it assumes the form represented in Figure 28, when it may be removed from the flame; and having been scratched with a file across its narrowest part, is gently broken asunder (Fig. 29). The sharp edges are then held in the flame until they are round and uniform (Fig. 30); after which the other end may be worked in the same way, only making it rather more tapering and pointed.

Fig. 27.

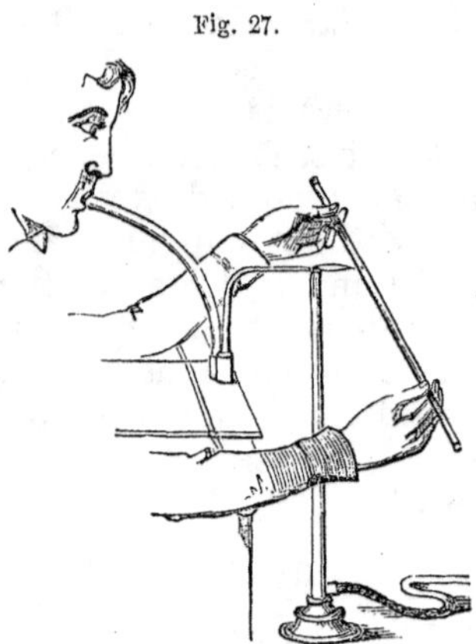

86. *Join together two rods of equal diameter.* For this purpose, take two short pieces of rod, the extremities of which are smooth and

Fig. 28.

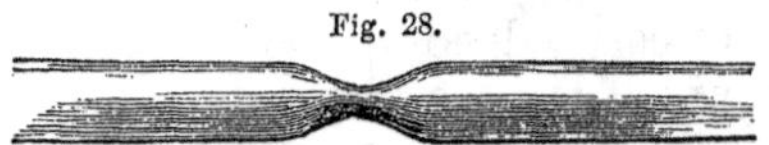

flat, and hold the ends which are to be united in the blowpipe flame until partial fusion takes place. Then with a steady hand bring them together, observing that

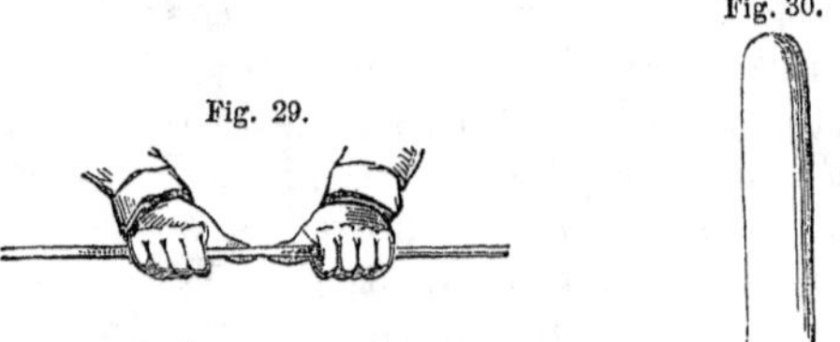

Fig. 29. Fig. 30.

the edges of both coincide, and press them gently, so as to cause them to cohere perfectly together. Keep the newly-formed joint in the flame for some minutes, turning it constantly round, and alternately pulling and pushing, in order to weld the two pieces firmly together.

When this is properly done, the rod is as strong at the junction as in any other part, but a slight inequality will always be visible, however neatly the operation may have been performed.

Fig. 31.

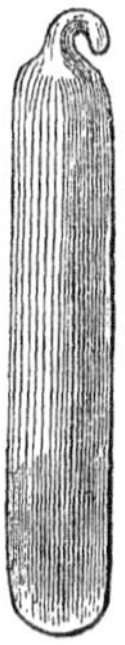

Specific-gravity glass.

87. *Make a specific-gravity glass* of the form and size shown in Fig. 31 (149).

88. *Make a small syphon tube.* Take a piece of tubing ten or twelve inches long, and a fourth or a third of an inch in diameter, and hold it diagonally in the flame of a gas or spirit-lamp, turning it constantly round, and by gently moving it up and down in the flame, heating two or three inches of the central part of the tube. When the glass begins to soften, apply a gentle pressure with both hands, so as to bend it slowly, and continue to do so until it has assumed the form shown in Fig. 32. If the tube is too strongly heated, or if the pressure be too strongly and suddenly applied, the bend, instead of being round and uniform, will be abrupt and wrinkled, in which case it is very liable to crack, either spontaneously, or when exposed to slight variations of temperature. The extremities of the tube must now be rounded off by being heated to redness for a moment in the flame of the blowpipe. When the glass operated on is at all thick, or of an unequal form, some care is necessary in *annealing*, or gradually cooling it; this may be effected by removing it slowly from the flame, and then laying it across a piece of tube, so that the hot part does not touch any cold substance, and covering it loosely with paper, to prevent too rapid cooling by radiation.

Fig. 32.

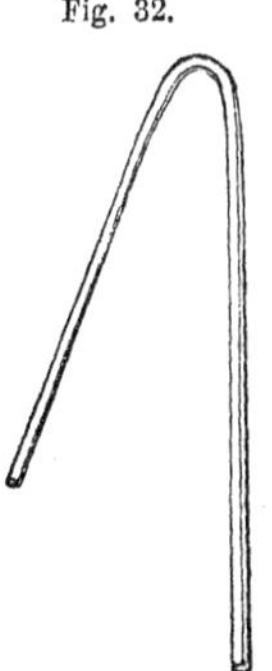

Syphon tube.

89. *Make a few test-tubes.* A piece of tube may be taken about half or five-eighths of an inch in diameter, and eight or ten inches long, which will serve for two test-tubes.

The central portion must be heated in the manner described for heating glass rod (85), and gradually drawn out, the tube being constantly turned round, when it will assume the form shown in Fig. 33. The heat should now be applied to the part of the tube marked *a*, and the other piece gradually drawn out, care being taken not to fuse the thin thread of glass that is formed,

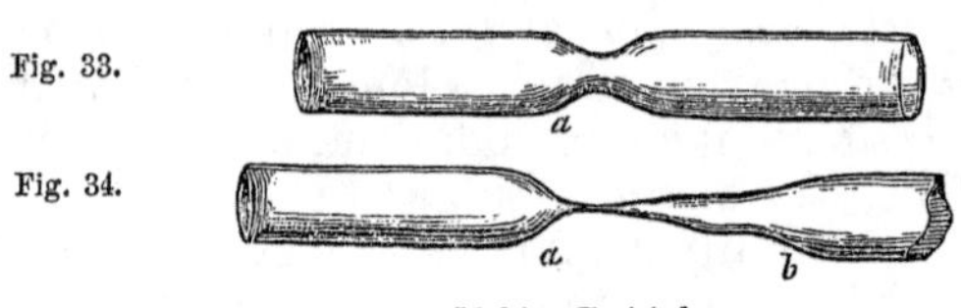

Fig. 33.
Fig. 34.
Making Test-tubes.

and which connects the two parts of the tube (Fig. 34), until the base of the tube has become round and uniform; when this is the case, and the connecting thread has become very thin, the heat may be applied to the point where it joins the tube, when it will instantly fuse and separate, leaving the tube in an almost finished state. There will generally be found at the bottom of the tube, however, a small lump, more or less distinct, formed by a portion of the thread having fused into it; to remove this, again heat the round end red-hot for a short time, until the lump disappears. On removing the tube from the flame, blow air gently into it, for the purpose of swelling out the bottom to its previous round form (Fig. 35), as it usually collapses and flattens while in a state of fusion.

Fig. 35.

90. The other portion of the tube may now be finished in a similar way, by applying heat to the point *b*, and drawing off the irregular termination until the thread of glass is sufficiently attenuated to be removed.

When it is required to make a test-tube of a piece of tubing only long enough for one, all that is necessary is to melt on to one end another piece of waste tubing or rod, to serve as a handle, after which the end may be drawn off, as in the former case.

91. To complete the tube, the open end must be spread out a little, as shown in Fig. 36, so as to form a kind of border. This is done by softening the end in the blowpipe flame, and then, by means of a thick iron

wire, or the smooth end of a file (which should be previously heated by being held in the flame), introduced and carried round the opening, the edge is uniformly pressed outwards.

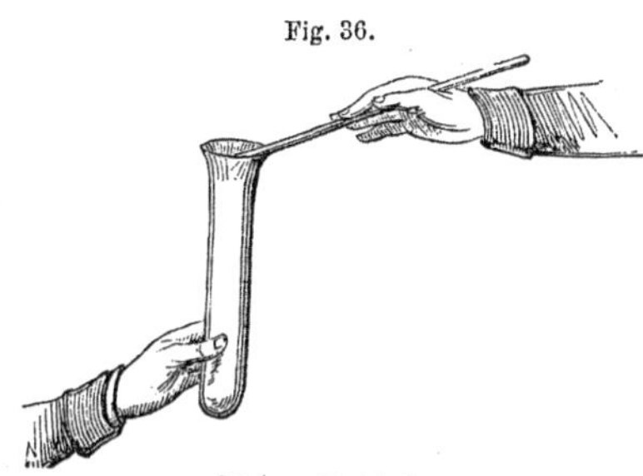

Fig. 36.

Making Test-tubes.

92. *Cement together two tubes of equal diameter.* This is done in a similar manner to that already described in the case of rods (86). It requires, however, more care and dexterity to maintain the tube of nearly uniform thickness at the point of junction, as it is liable to collapse and become irregular in form. When it does so, one end of the tube should be stopped up with a bit of cork or by hermetically sealing, and while the junction is in a state of semi-fusion, air should be gently blown into the tube: in this way it may be brought again into a proper form. When the glass is thin, the edges which are to be united may be spread out a little, as shown in Fig. 37, by means of a heated wire or file (91), when the joint will be stronger than it would otherwise be.

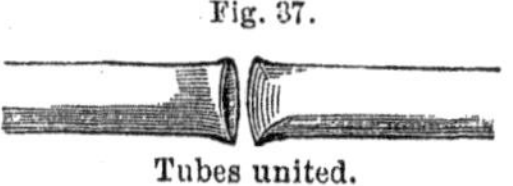

Fig. 37.

Tubes united.

93. *Cement together two tubes of unequal diameter.* When it is required to join a narrow tube to a wider one, it is necessary to draw out the latter in the blowpipe flame until a portion of it is contracted to the diameter of the former (85); then with a file it is divided at that point of equal diameter, and cemented to the smaller tube in the same way as in the previous case. Sometimes, when the glass is thin, it is advisable to widen the extremity of the smaller tube, so as to overlap the other (Fig. 38), which is readily done by means of an iron wire (91).

In this operation, it is always advisable to maintain the junction in the flame for some little time, to allow of the complete amalgamation of the two portions of glass; and as the tendency to collapse is greater the longer it is fused, it will generally be found necessary to blow it out slightly, as recommended in (92).

Fig. 39.

Funnel tube.

In this way, some small funnels may be made (Fig. 39).

Fig. 38.

Funnel tube.

94. *Prepare tubes for a washing bottle.* The tubes required for this purpose are of the form shown in Fig. 40, the upper end of the longer one being drawn out so as to leave only a small aperture.

When the bottle is prepared and filled with water, a small stream of water may be forced through this tube by blowing air down the shorter one; it is of great service in washing precipitates on a filter, and for many other purposes (169).

Fig. 40.

Washing bottle.

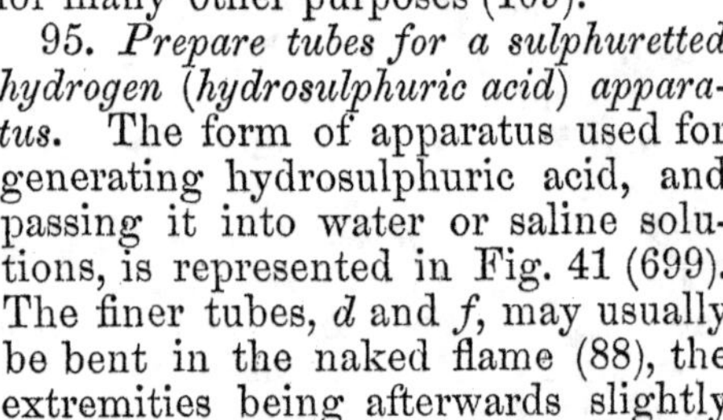

95. *Prepare tubes for a sulphuretted hydrogen (hydrosulphuric acid) apparatus.* The form of apparatus used for generating hydrosulphuric acid, and passing it into water or saline solutions, is represented in Fig. 41 (699). The finer tubes, *d* and *f*, may usually be bent in the naked flame (88), the extremities being afterwards slightly fused with the blowpipe in order to round off the sharp edges; and care must be taken that the wider tube *e* is of sufficient calibre to admit of the tube *d* passing freely down it.

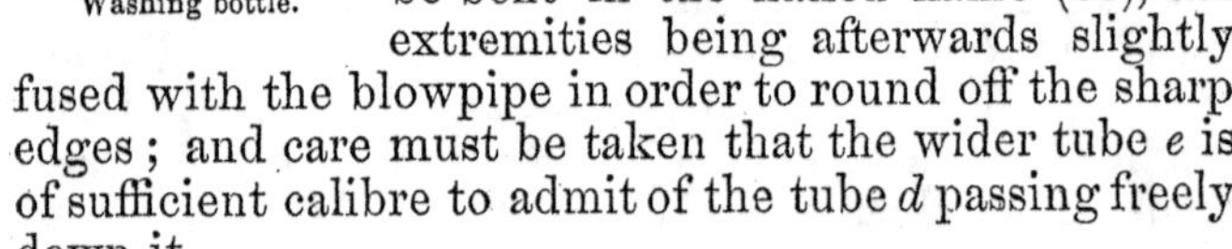

96. *Attach a narrow tube at right angles to a wider one.* Heat the wider tube to redness at the point where the junction is to be made; and by means of a bit of waste rod or tubing *c* (Fig. 42), draw it out, when it will assume the form represented in the figure: then with a sharp file remove the portion *c* at the point *b*, and fuse to the

projecting piece thus left the smaller tube, in the manner described in (93).

Fig. 41.

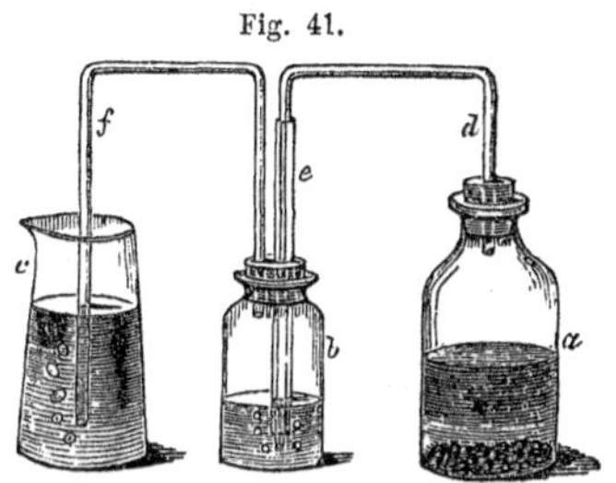


Sulphuretted Hydrogen Apparatus.

97. *Blow some small bulbs.* When it is required to blow a bulb at the end of a tube, the extremity should be closed as in making a test-tube (89); if the glass is tolerably thick, and the bulb to be blown not large, all that is necessary is to heat the end for about half an inch as strongly as possible; and then, having removed it from the flame, and holding it horizontally in the hands (Fig. 44), to blow air into it until the pressure forces the softened glass to expand, which it will do in the form of a round bulb if the heat has been properly applied, and the tube be kept constantly turned round while in the hands. This latter precaution is absolutely necessary, as the softened glass would otherwise bend with its own weight in one direction, thus destroying the proper form of the tube.

Fig. 43. Fig. 42.

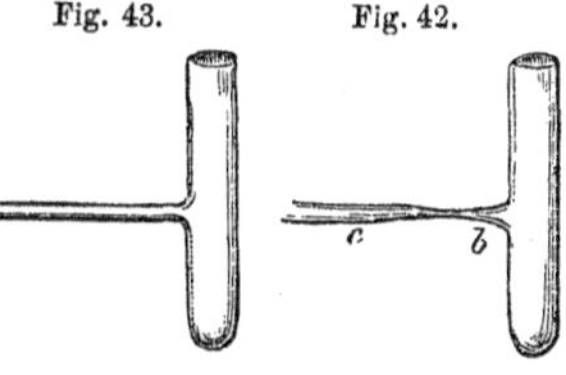


Fig. 44.

98. *Seal a few tubes hermetically at both ends.* This is an operation of very frequent use in the laboratory, as it

furnishes the most convenient and efficient means of preserving small specimens of many rare substances, especially such as are volatile.

The tube is first sealed at one end, precisely as if it were intended for a test-tube (89); the liquid or other substance for which it is designed is then introduced, as soon as the tube is quite cold, care being taken that the upper part of the tube is not wetted or soiled. The flame of the blowpipe is now directed to the portion of the tube a little above that intended for the sealed end, and when sufficiently soft it is drawn out to a capillary tube, and allowed to cool: it may afterwards be sealed by fusing the lower part of the capillary tube *a* (Fig. 45), by momentary contact with the flame.

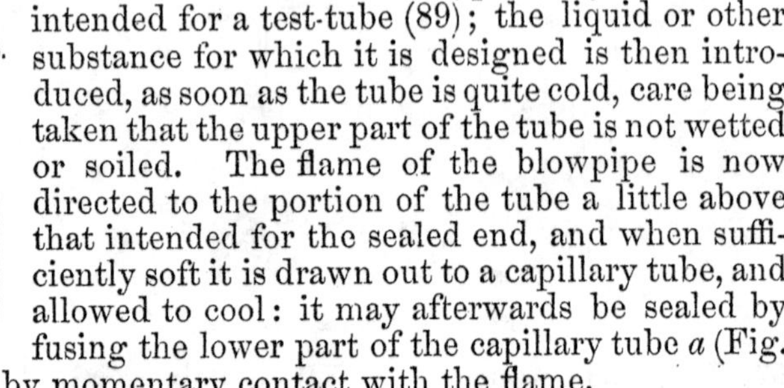

Fig. 45.

In this way seal a little sulphur in a tube without melting or volatilizing any of it, the sulphur being within an inch and a half of the upper end.

99. *Seal some water hermetically in a tube.* Having sealed the tube at one end, while it is cooling take another piece of tubing, which may be eight or nine inches long, and a quarter of an inch in diameter, and draw it out in the blowpipe flame; then divide it in the thin part by means of a file, when it will have the form shown in Fig. 46; and when the sharp edges have been rounded off in the blowpipe flame, may be used

Fig. 46.

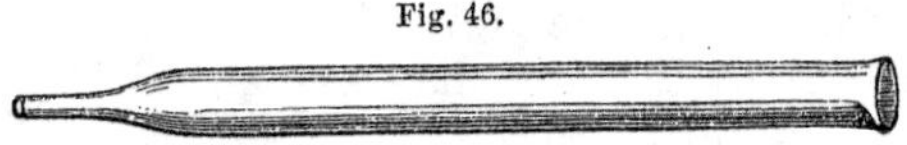

as a pipette for introducing a little water into the sealed tube without wetting its sides.

Then draw out the capillary neck (98), and when cold seal it as before, leaving not more than the space of an inch between the upper end and the surface of the water.

# CHAPTER V.

## EXPERIMENTS WITH THE MOUTH BLOWPIPE.

100. Before proceeding to any blowpipe experiments, it is necessary to acquire the knack of keeping up a constant and unintermitting blast of air from the mouth, as without this it is impossible to raise the heat to a sufficient degree of intensity. The habit is readily acquired, and when once attained, the mouth and lungs will be found to do their work almost mechanically, without any sustained effort on the part of the operator.

101. The learner may first observe that on closing the lips he can still without any difficulty breathe through the nostrils: let him now distend the cheeks with air from the lungs, and he will find that on closing the communication between the mouth and throat he can breathe through the nostrils for a length of time, still keeping the cheeks distended. He may next introduce the mouth-piece of the blowpipe between his lips, and having puffed out his cheeks with air from the lungs, and again closed the communication between the mouth and throat, let him breathe freely through the nostrils, at the same time allowing the distended cheeks to force a current of air through the blowpipe. When the stock of air in the mouth is nearly exhausted, a fresh supply is sent up from the lungs, when the cheeks, again distended, will by their elasticity keep up a current of air through the blowpipe, while the operator breathes through the nostrils as before.

The cheeks thus play the part of an elastic bag, with a valve opening inwards, which, if connected with the blowpipe, and distended with air, would force air through it as long as the tension of its stretched sides exerted sufficient pressure.

102. Seal a few tubes for the following experiments (Fig. 47). The tubing employed for this purpose should be about a quarter of an inch in diameter, and it may be cut into pieces about five inches long, each of which

will serve for two tubes. The sealing should be effected in the manner already described (89), and the same care relative to the deoxidizing flame is necessary, as when the water blowpipe is used (84).

Fig. 47.

103. Heat a small fragment of wood or paper in a tube, and observe that it blackens like all organic substances.[1] This blackening or charring is owing to the decomposition of the lignine, which consists of $C_{12}H_{10}O_{10}$; when exposed to a high temperature the hydrogen, and oxygen, with a portion of the carbon, pass off in the form of acetic or pyroligneous acid ($HO,C_4H O_3$) and tarry matter, with other volatile compounds, leaving behind a carbonaceous residue. The acid character of the vapor may be seen by introducing a strip of moistened litmus-paper into the upper part of the tube while the decomposition is going on, when it will be speedily reddened.

104. Treat a fragment of horn ($C_{48}H_{39}N_7O_{17}$), or isinglass ($C_{96}H_{82}N_{15}O_{36}$), similarly, in another tube:[2] observe the character of the carbonaceous residue, and introduce a bit of yellow turmeric paper, which will be turned brown, showing that the vapor is alkaline; this is owing to the presence of ammonia ($NH_3$), which is almost invariably produced when an organic compound containing nitrogen is decomposed by heat. The odor of the fumes should also be noticed, and contrasted with those formed in the last experiment.

105. Heat a little gypsum or sulphate of lime (CaO, $SO_3$+2Aq) in a tube, and note whether it undergoes any change. It parts with the two equivalents of

[1] In this and most of the following experiments, especially when the substance operated on is of a deleterious or poisonous nature, the quantity used should not exceed a pin's head in size.

[2] When a tube is at all soiled in an experiment, it is unfit for further use.

water of crystallization, which condense in the upper part of the tube.

106. Treat a crystal of sulphate of iron ($FeO,SO_3+7Aq$) in a similar manner, observing the successive changes which are produced, and examine the liquid which condenses in the upper part of the tube, with litmus-paper.

When first heated, six equivalents of water are expelled, leaving a whitish powder, which consists of the sulphate with one equivalent of water ($FeO,SO_3,HO$). On continuing the heat, the sulphuric acid is volatilized, a portion of it being decomposed by the protoxide of iron, which is converted into peroxide by the oxygen derived from the acid.

$$2(FeO,SO_3)=\mathit{SO_3}+SO_2+Fe_2O_3.$$

107. Repeat the experiment, using sulphate of potash ($KO,SO_3$) instead of sulphate of iron; the decrepitation is owing to the escape of a little water, which is mechanically lodged between the plates of the crystals. The salt undergoes no further change.

108. Sublime a little calomel ($HgCl$), and corrosive sublimate ($HgCl_2$) in two separate tubes, and note the different appearances which are presented in both cases.

109. Heat a little red oxide of mercury ($HgO_2$) in a tube; observe the rapid change which it undergoes, and the minute globules of metallic mercury which condense in the upper part of the tube. If a glowing match be introduced while the decomposition is going on, it will indicate, by its vivid combustion, the presence of free oxygen, especially if the open end of the tube be loosely closed with the finger, to retard the escape of the disengaged oxygen.

$$HgO_2=\mathit{Hg}+2O.$$

110. Repeat the experiment with some red oxide of lead ($Pb_3O_4$), and observe in what respects the results differ from the last. The yellowish residue which is left is protoxide of lead or litharge ($PbO$), one-fourth of the oxygen being expelled.

$$Pb_3O_4=3PbO+O.$$

111. Heat a little arsenious acid ($AsO_3$) in a tube, and

observe closely the characters of the crystalline sublimate (302).

112. Mix together equal portions of nitre ($KO,NO_5$) and bisulphate of potash ($KO,HO,2SO_3$), and heat the mixture in a tube; test the nitrous vapor which is given off, with litmus-paper, and endeavor to account for its formation.

113. Heat a mixture of pounded fluor spar ($CaF$) and bisulphate of potash ($KO,HO,2SO_3$) in a glass tube. The corrosive action on the glass is owing to the formation of hydrofluoric acid ($HF$).

$$KO,HO,2SO_3+CaF=HF+CaO,SO_3+KO,SO_3.$$

114. Mix a little iodide of lead ($PbI$) with bisulphate of potash, and heat the mixture in a tube: the beautiful violet-colored vapor which rises and condenses in the upper part of the tube is iodine.

$$PbI+KO,HO,2SO_3=PbO,SO_3+KO,SO_3+H+I.$$

115. Fuse a little phosphate of lead ($3PbO,PO_5$) on charcoal, and observe the semi-transparent crystalline appearance of the bead on cooling (412).[1]

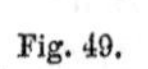

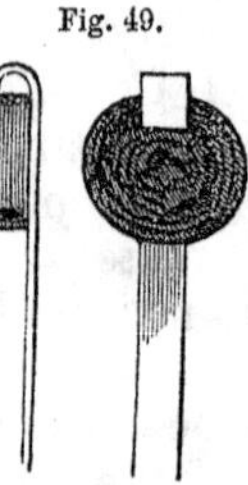

Fig. 49.

Charcoal-holder.

116. Heat a little oxide of zinc ($ZnO$) on charcoal; observe that it assumes a yellow color when heated, but becomes white again on cooling.

117. Notice the change of color that ensues when chromate of lead ($PbO,CrO_3$) is gently heated, and observe whether the yellow color returns on cooling.

118. Repeat the experiment with red oxide of mercury ($HgO_2$), taking care that the heat is not raised so high as to cause decomposition (109).

119. Mix together a little chalk ($CaO,CO_2$) and charcoal, and ignite the mixture in a tube; carbonic oxide

[1] When charcoal is used as a support in blowpipe experiments, it should be cut into slices about the third of an inch in thickness, having a small cavity scooped out with the point of a knife, in which to lodge the substance to be heated. The charcoal may be conveniently held during the experiment in a loop of tin plate, in the manner shown in Fig. 49.

gas is given off, which, if formed in sufficient quantity, will burn with a blue flame.

$$CaO,CO_2+C=CaO+2CO.$$

120. Heat a small crystal of carbonate or any other salt of soda on platinum wire (which should be fused into a glass handle, and bent at the end, as shown in Fig. 48), and observe the intense yellow color it communicates to the blowpipe flame. Then wash the wire, and compare its action on the flame with that caused by the soda.

Fig. 48.

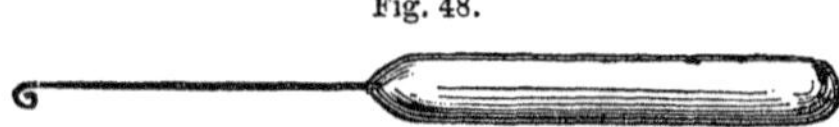

121. Repeat the experiment, using nitrate of strontia ($SrO,NO_5$) instead of the soda: the color of the flame will become crimson.

122. Heat a little chalk or marble ($CaO,CO_2$) on charcoal, and note the dazzling white light which is produced, showing that the illuminating power of flame is not dependent only on the degree of heat, but on the presence of some solid matter in the flame;[1] since the blowpipe flame, which heats it, and which is of course at least as hot as the lime, emits scarcely any perceptible light.

During the ignition, the carbonate of lime is decomposed, and caustic lime ($CaO$) is left, the alkaline nature of which may be shown by placing a fragment of it, after ignition, on moistened turmeric paper, which will become brown at the point of contact.

123. A piece of alumina ($Al_2O_3$) or alum ($Al_2O_3,3SO_3+KO,SO_3+24Aq$) ignited in the flame, radiates a faint bluish light.

124. Dip a glass rod in a solution of nitrate of cobalt ($CoO,NO_5$), and moisten a small crystal of alum with it; then ignite it on charcoal for a few minutes, and observe the beautiful blue color which it assumes. This is a highly characteristic test for alumina.

125. Repeat the experiment, with sulphate of magne-

[1] See Daniell's Chemical Philosophy, p. 361.

6

sia ($MgO,SO_3+7Aq$), which, when ignited with nitrate of cobalt, gradually assumes a pale rose color.

126. A salt of zinc, as the sulphate ($ZnO,SO_3+7Aq$), when similarly treated, becomes green.

It is easy therefore to distinguish between alumina, magnesia, and zinc, in this simple manner.

127. Heat a fragment of tin in the deoxidizing flame until it fuses into a bright metallic globule; when white hot, throw it on the table, when it will divide into numerous small globules, which run rapidly about, burning with a white light, and leaving behind them white trains of oxide ($SnO_2$).

128. Heat another fragment of tin, and keep it fused and bright in the deoxidizing flame for two or three minutes; then oxidize it in the outer flame, and again reduce it to the metallic state.

129. Heat a little acetate of lead ($PbO,C_4H_3O_3+3Aq$) on charcoal; observe first the liberation of acetic acid ($HO,C_4H_3O_3$) and the deposition of a portion of the carbon; and on a further application of heat, the oxide of lead first deposited is reduced to the metallic state, especially when it is kept in the deoxidizing flame. The yellow ring which surrounds the metallic bead is protoxide of lead ($PbO$).

130. Reduce oxide of bismuth ($Bi_2O_3$) in the same way: compare the beads of the different metals thus obtained, as to outward appearance, crystalline structure, malleability, &c.

131. Heat a small crystal of sulphate of copper ($CuO, SO_3+5Aq$) in the reducing flame on charcoal, and observe the successive changes which it undergoes; first into black oxide ($CuO$), and ultimately into a bead of metallic copper. Hammer out the globule, so as to render visible its peculiar color.

132. Mix together a little sulphate of baryta ($BaO, SO_3$) and charcoal in a mortar, and fold a small quantity of the mixture under one corner of a slip of platinum foil (Fig. 50). Heat it strongly in the blowpipe flame, and when the ignited mixture is cool again, put it into a small tube, and treat it with a drop or two of dilute hydrochloric acid (*HCl*). Observe the effervescence

caused by the escape of hydro-sulphuric acid gas (HS), which may be recognized by its peculiar odor, and by a piece of paper moistened with a solution of acetate of lead, which is instantly blackened by it (438). In this experiment the sulphate of baryta is deoxidized by the charcoal, becoming sulphide of barium (BaS), which, when acted on by hydrochloric acid, is decomposed, and hydrosulphuric acid liberated.

Fig. 50.

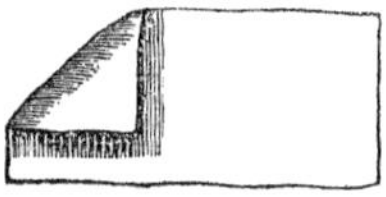

*Decomposition during ignition.*

$$BaO,SO_3+2C=BaS+2CO_2.$$

*Decomposition caused by the hydrochloric acid.*

$$BaS+HCl=BaCl+HS.$$

133. Sublime a little sulphur in a small tube open at both ends; while in the state of vapor in contact with the atmospheric oxygen, it becomes converted into sulphurous acid ($SO_2$), the presence of which may be shown by its property of reddening litmus paper when moistened, and bleaching it when dry: its smell also is well known and characteristic.

134. Heat a small quantity of sulphide of antimony ($SbS_3$) in an open tube (Fig. 51); observe the formation of oxide of antimony ($SbO_3$) which appears as white fumes, and test for the presence of sulphurous acid ($SO_2$) as in the last experiment. Here the oxygen of the air has oxidized both the sulphur and the metal.

Fig. 51.

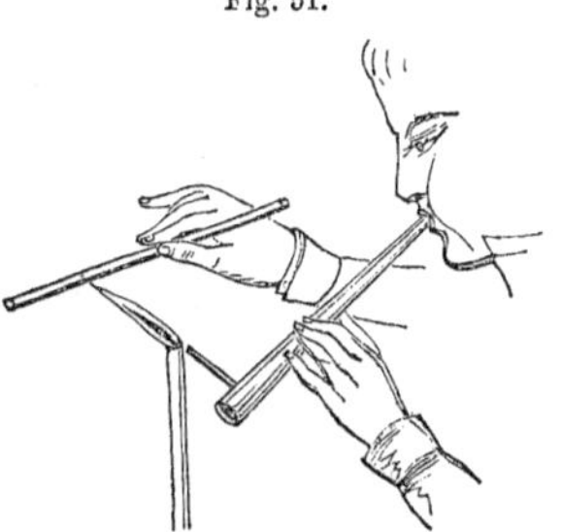

135. Scoop out a cavity in a piece of charcoal *a* (Fig. 52), and nearly fill it with a paste made of phosphate of lime ($8CaO,3PO_5$) and water, *b*; dry it on the sandbath, and when quite dry, place a fragment of lead upon it. Expose it to the oxidizing flame, and observe that the oxide of lead (PbO) as it is formed, is absorbed by the porous

Fig. 52.

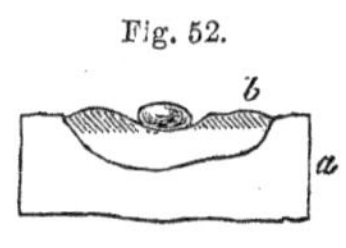

phosphate of lime, while any silver which may be present, is left unoxidized, as a small metallic bead. This process is called *cupellation.*

136. Fuse a little carbonate of soda ($NaO,CO_2$) on charcoal, and observe that it is absorbed, owing to the capillary attraction of the porous charcoal.

137. Make a bead of glass, by fusing a mixture of carbonate of soda and silica ($SiO_3$) (427).

138. Add a little sulphate of lime to the bead formed in the last experiment: heat it strongly in the deoxidizing flame, and remark the yellow color which it assumes, owing to the formation of sulphide of sodium ($NaS$).

$$NaO,SiO_3 + CaO,SO_3 = CaO,SiO_3 + 4O + NaS.$$

139. Mix a little black oxide of manganese ($MnO_2$) with carbonate of soda ($NaO,CO_2$), and fuse it on platinum wire: remark the characteristic green color which is produced (267).

Fig. 53.

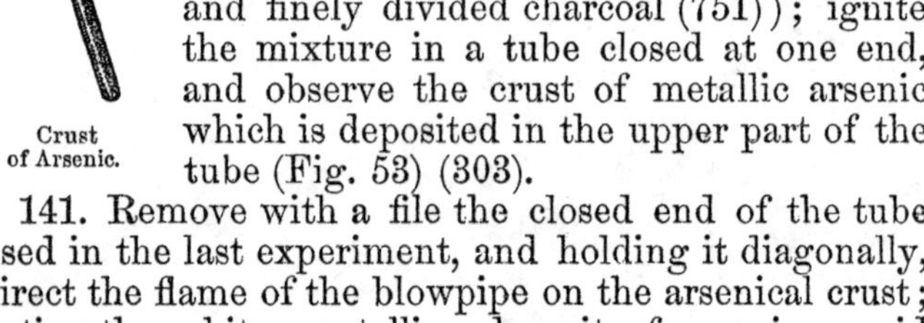


Crust of Arsenic.

140. Mix together a little arsenious acid ($AsO_3$) (a quantity not greater in bulk than a small pin's head) (301) and black flux (which is a mixture of carbonate of potash and finely divided charcoal (751)); ignite the mixture in a tube closed at one end, and observe the crust of metallic arsenic which is deposited in the upper part of the tube (Fig. 53) (303).

141. Remove with a file the closed end of the tube used in the last experiment, and holding it diagonally, direct the flame of the blowpipe on the arsenical crust; notice the white crystalline deposit of arsenious acid ($AsO_3$) which condenses in the cool part of the tube, and examine with a lens the beautiful octohedral crystals of which it is composed (Fig. 54).

142. Place a little calomel ($HgCl$) or corrosive sublimate ($HgCl_2$) at the bottom of a tube, and cover it for about half an inch with dry carbonate of soda ($NaO, CO_2$); and make the upper portion of the salt quite hot, and carrying the heat downwards sublime the calomel through it. Metallic mercury is deposited in

the form of minute globules in the upper part of the tube (336).

$HgCl+NaO,CO_2=NaCl+Hg+O+CO_2$.

143. Fuse a little borax ($NaO,2BO_3+10Aq$) on a platinum wire, and observe the color given to the bead by salt, iron, cobalt, copper, lead, manganese, &c., both in the oxidizing and reducing flame.

Fig. 54.

Arsenious Acid.

144. Repeat the same series of experiments, using carbonate of soda ($NaO,CO_2$), and afterwards microcosmic salt ($NaO,NH_4O,HO,PO_5+8Aq$), instead of borax, and observe in what respects the results differ from each other.

---

# CHAPTER VI.

## SPECIFIC GRAVITY.

By *specific gravity* is meant the comparative weights of equal bulks of various kinds of matter. It has been found convenient to compare the specific gravities of all solids and liquids with that of water, which is reckoned as 1·000 or 1000. The specific gravity of substances heavier than water is consequently represented by a higher number, and those which are lighter by a lower number, than 1·000; that of lead, for instance, which is more than eleven times heavier than water, is represented by the number 11·35; while that of ether, which is considerably lighter than water, is represented by the number 0.724.[1]

### SECTION I.

*Specific gravity of solids heavier than water.*

145. When the substance is solid and insoluble in water, its specific gravity may be ascertained in the following manner. Weigh it first in air, taking care

[1] See Fownes's Manual of Chemistry, p. 3.

Fig. 55.

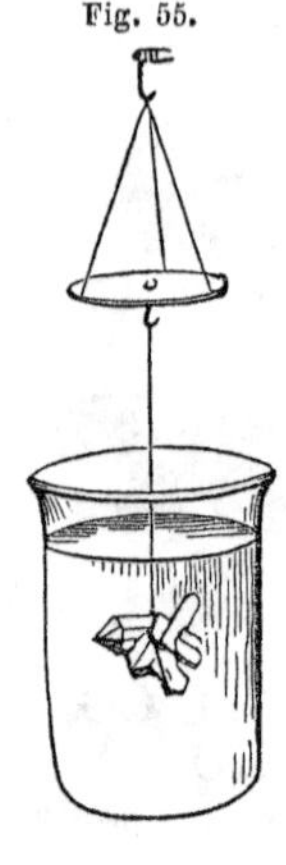

to remove any dust or loosely-adhering particles. Then suspend it by means of a horsehair, from a hook attached to the scale-pan, making a small loop at one end of the hair, passing the other end through it, and inclosing the substance in the noose. Thus suspended, it is immersed in water (Fig. 55), and care should be taken that it is covered on all sides by at least half an inch of water. Small bubbles of air frequently adhere to the surface, and these must be brushed off with a feather or camel-hair pencil, as they would tend to buoy it up, and cause the specific gravity to appear too low.

The results may be noted down as follow:

Weight of the substance in air . . . =
" " in water . . = ______
Loss . . = ______

which number represents the weight of an equal bulk of water. Then by dividing the weight in air by the loss, or the weight of an equal bulk of water, the specific gravity is ascertained.

$$\frac{\text{Weight in air}}{\text{Loss}} = \text{specific gravity.}$$

In this way determine the specific gravity of some of the following substances:—marble, amber, iron-pyrites, sulphate of baryta, jet, lead, zinc, glass, and agate.[1]

Fig. 56.

## SECTION II.

*Specific gravity of solids lighter than water.*

146. If the solid be lighter than water, as cork, a slight modification of the above process is necessary.

Weigh the substance first in air; then

[1] The following are the specific gravities of these substances, some of which, however, vary considerably. Marble 2·70; amber 1·08; iron-pyrites 4·90; sulphate of baryta 4·47; jet 1·30; lead 11·35; zinc 7·00; flint glass 3·30; and agate 2·60.

select a piece of lead of sufficient size to sink the light body in water when attached to it, and weigh it (the lead) in water, suspending it by means of a hair loop, as before. If now the light substance be inclosed in the same loop with the lead (Fig. 56), and immersed in water, it will be found that they will together weigh less than the lead did alone, owing to the buoyancy of the lighter body; and this difference, when added to the weight of the body in air, is equal to the weight of a corresponding bulk of water.

The results may be thus recorded:

| | | |
|---|---|---|
| Weight of body in air . . . . . . . . | = | ———— |
| Weight of lead alone in water . . . . . . | = | |
| Weight of lead with body attached, in water | = | ———— |
| Difference . . . . . . . | = | |
| Add weight of body in air . . . . . . . | = | ———— |
| Weight of an equal bulk of water . . . . | = | ==== |

Having thus obtained the weight of the body in air, and the weight of an equal bulk of water, the specific gravity is calculated as before.

$$\frac{\text{Weight in air}}{\text{Weight of equal bulk of water}} = \text{Specific gravity.}$$

In this way ascertain the specific gravity of wood, cork, and charcoal.[1]

## SECTION III.

*Specific gravity of insoluble powders.*

147. When the substance, whose specific gravity we wish to determine, is in the form of powder or even small lumps, it is clear that some other method must be adopted than those just described. The following is the most simple, and for common purposes, sufficiently accurate. Counterpoise[2] a small bottle furnished with

[1] The specific gravity of these substances varies considerably, according to the degree of porosity; the following may be considered as the usual average: wood (beech) 0·85; cork 0·24; and charcoal 0·2 to 0·5.

[2] This is done by putting shot or strips of lead in a pill-box, which, when placed in the opposite scale, are adjusted until their weight is equal to that of the bottle.

a stopper; then fill it completely with distilled water, close it with the stopper, taking care that no bubbles of air are left in, and weigh to determine the quantity of water it contains.[1] Having done this, empty the bottle, and dry the inside either with a cloth, or with fragments of filtering paper.

It must now be filled about two-thirds full of the powder to be examined, again weighed, and the bottle then filled cautiously with water, care being taken that all air-bubbles are expelled, and that none of the powder is washed out. Again weigh.

From the data thus obtained, the specific gravity may be calculated as follows:

| | |
|---|---|
| Weight of the powder and water . . . = | |
| Weight of the powder alone . . . . . = | |
| Difference = weight of water left in the bottle | |
| Weight of bottle full of water . . . . = | |
| Water left in the bottle after the powder was added . . . . . . . . . = | |
| Weight of water displaced by, and equal in bulk to, the powder . . . . = | |

Then as before:

$$\frac{\text{Weight of the powder}}{\text{Weight of water displaced}} = \text{Specific gravity.}$$

In this way ascertain the specific gravity of sand, pounded glass, and shot.[2]

## SECTION IV.

### *Specific gravity of liquids.*

148. With a bottle similar to that used in the last experiment, the specific gravity of liquids may be readily determined. As the space occupied by a given weight of liquid varies with the temperature, or, in other words, as the weight of a given volume of any liquid is greater or less as the temperature is lower or

[1] Bottles may be purchased which are made to contain exactly 1000 grains of distilled water.

[2] The specific gravity of sand is about 2·60; flint glass 3·30; and shot 11·35.

higher, it is necessary to observe that the temperature of the liquid during the experiment does not vary much from 62°, which is usually taken as the standard. For the same reason the bottle should not be touched by the warm hand during the experiment, as otherwise the heat would cause the liquid to expand, and become specifically lighter; this may be avoided by interposing a linen cloth between the hand and the glass.

Fig. 57.

Specific-gravity glass.

Counterpoise the bottle, and weigh it full of distilled water; then, by filling it successively with other liquids, weighing, and comparing the different weights with that of water, the volume of liquid being always the same, the specific gravity is obtained by proportion, thus:

Weight of bottle full of water : 1·000 : : Weight of liquid : Specific gravity.

Care must be taken to clean the bottle thoroughly after each experiment, by washing it first with distilled water, and then with a little of the liquid whose density is to be ascertained.

Some of the following may be taken for practice:—Alcohol, solutions of chloride of sodium, sulphate of magnesia, alum, carbonate of soda, sulphate of lime, sulphate of soda, bicarbonate of soda, sulphate of copper, nitrate of potash, sulphate of zinc, and cream of tartar.

Fig. 58.

149. The specific gravity of liquids may also be determined by another process, which, though not capable of so much accuracy as the last, is frequently useful when the specific gravity bottle is not at hand.

Take a piece of solid glass rod, about the size of the figure (Fig. 57), with one end drawn out and turned in the blowpipe flame. Weigh it first in air and then in water, suspending it with a hair-loop (Fig. 58). Then, having wiped it dry between each experiment, weigh it successively in the liquids, the specific gravities of which are to be determined. The difference between the weight of the glass in air and in the liquid, representing in each case the weight of a volume of the liquid equal to that of the glass, and knowing the weight of a similar volume of water, the specific gravity may be known by simple calculation.

Thus:

| | |
|---|---|
| Weight of glass in air . . . . . . . . . . | |
| Weight of glass in liquid . . . . . . . . | |
| Loss . . . . . . . . . . . . | |

which is the weight of an equal volume of the liquid.

Then by proportion,—

Weight of equal volume of water. : 1·000 :: Weight of equal volume of liquid : Specific gravity of the liquid.

Determine in this way the specific gravities of some of the solutions already mentioned, and compare the results with those obtained with the specific gravity bottle.

---

## CHAPTER VII.

### HEATING SUBSTANCES IN GASES.

#### SECTION I.

*Reduction of metallic oxides by hydrogen.*

150. A LARGE number of the metallic oxides are decomposed and reduced to the metallic state when heated in an atmosphere of dry hydrogen gas; and from the facility with which the operation may be performed, and the accurate results it gives when carefully conducted, it is frequently employed in estimating the quantity of oxygen present in oxidized compounds.

151. The apparatus which is required for the purpose is shown in the figure. (Fig. 59.) The bottle *a* is charged with zinc and dilute sulphuric acid to generate the hydrogen, which is dried while passing over fragments of chloride of calcium in the tube *e;* the gas then passes into the bulb-tube *h*, which contains the oxide to be reduced, the bulb being heated by the lamp placed beneath.

152. Take a piece of tubing *e*, about twelve or fifteen inches long, and half an inch internal diameter, and having slightly fused the cut edges in the blowpipe

Fig. 59.

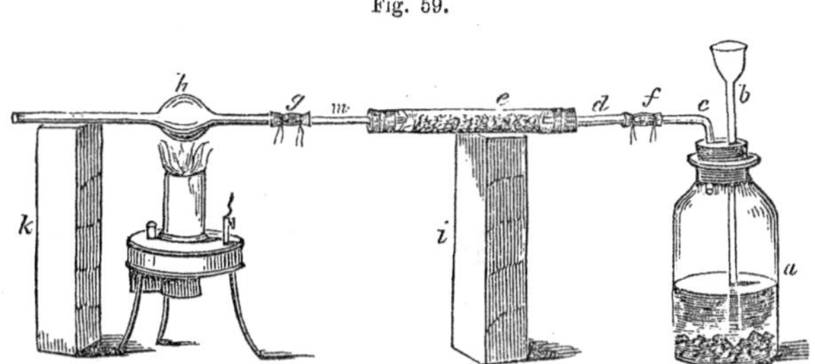


Reduction of Oxide of Copper.

flame (85), adapt a cork to each end: then, with a cork-borer or round file perforate the corks so as to receive the small tubes *d* and *m*. When the tube is of such a diameter as cannot be exactly matched by any of the cork-borers in the set, the hole should be bored by a smaller one and afterwards enlarged by means of a round file, until it is of sufficient calibre to admit the tube, which must always fit perfectly tight. Remove one of the corks from the large tube, and push down to the other end a small loose bit of tow or cotton wool, and nearly fill it with fragments of chloride of calcium (734); put in another bit of tow (the use of which is to prevent any of the smaller fragments falling out), and again fix the cork and small tube.

Next adapt a cork to the bottle, which should have a tolerably wide neck, and bore in it two holes to fit the tubes *b* and *c*, which pass through it, the former reaching nearly to the bottom of the bottle, the latter passing only just through the cork. Put 300 grains of granulated zinc into the bottle, and fix the cork containing the tubes *b* and *c*.

153. In order to connect the different parts of the apparatus together, make two caoutchouc connectors *f* and *g*. This is done by loosely folding a piece of sheet caoutchouc about an inch and a half square, round a piece of rod or tubing of the same diameter as the tubes

which it is intended to join together, and cutting off with one stroke of a pair of sharp scissors the superfluous ends (Fig. 60); when this is properly done, the cut edges cohere, and when slightly pressed together by the thumb-nails, the junction becomes almost as strong as any other part of the tube. Care must be taken to

Fig. 60.

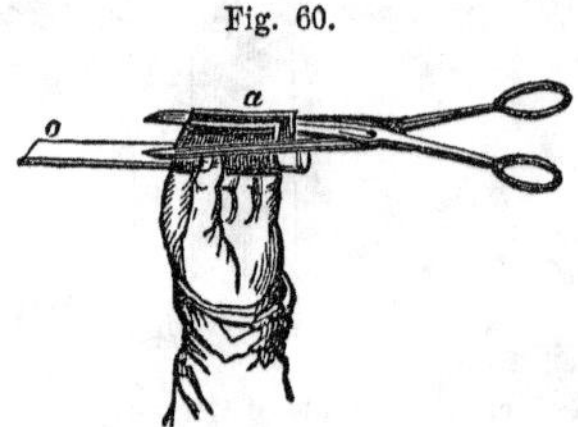


Making Caoutchouc Connectors.

avoid touching the newly-cut edges, as the least dirt or moisture upon them would prevent them cohering properly together. It is then carefully removed from the rod, and is ready for use.

154. Having made two of these connectors, weigh the bulb-tube accurately, and place in the bulb 20 or 30 grains of oxide of copper (CuO): again weigh, to ascertain the weight of oxide operated on, and connect the apparatus as shown in the figure. The caoutchouc tubes should be firmly tied round with strong twine or silk, which should be passed under and over, and tied at each half revolution to insure perfect tightness of the joint. The apparatus being thus arranged, fill the bottle *a* about one-third full of dilute sulphuric acid (consisting of one part by measure of the strong acid and eight parts of water), pouring it down the funnel-tube *b* (12); and when the gas has been coming over about five minutes, apply a gentle heat to the bulb, and gradually increase it as long as any water is formed.

$$CuO+H=Cu+HO.$$

It is necessary to observe the precaution of not applying the heat immediately, since the apparatus at first contains an explosive mixture of hydrogen and common air, which would, if heat were applied, be in great danger of exploding (17) and seriously injuring the operator; by allowing five minutes to elapse, however, the whole of the common air is expelled, and the bulb may be heated without danger.

155. When the decomposition appears to be complete, no fresh water being produced,[1] expel by heat any moisture that may have condensed in the cool end of the tube, remove the lamp, and allow the bulb-tube to cool; then disconnect the apparatus, and weigh the bulb containing the reduced metallic copper, the loss of weight indicating the quantity of oxygen that has been removed. Ascertain by calculation the percentage of oxygen in 100 parts of the oxide, and compare the experimental result with what is theoretically correct, the atomic weight of copper being thirty-two, that of oxygen eight, and that of the oxide forty.

## SECTION II.

### *Heating substances in an atmosphere of carbonic acid.*

156. It is sometimes required in analysis to separate two substances, one of which is volatile at a high temperature, and the other fixed, so that by merely heating the mixture, and weighing before and afterwards, the weight of each ingredient is determined. In some cases, however, it happens that the non-volatile body when heated in atmospheric air, combines with oxygen, forming a volatile compound; so that here it is necessary to conduct the operation in an atmosphere of some gas incapable of combining with it, as hydrogen or carbonic acid. For instance, in the analysis of gunpowder, which consists of a mixture of nitrate of potash ($KO,NO_5$), sulphur, and charcoal, the nitrate of potash is first dissolved out with water, and the insoluble residue, consisting of sulphur and charcoal, is heated in a current of hydrogen or carbonic acid, when the sulphur, being volatile, is expelled; whereas, if the mix-

[1] This is known by holding a piece of cold glass close to the opening at the end of the tube, and observing whether any moisture is condensed upon its surface; if not, it may be inferred that no water is coming off.

ture were to be heated in common atmospheric air, the carbon as well as the sulphur would disappear, since it would combine with oxygen, and become converted into carbonic acid ($CO_2$) which is a gas.

157. The apparatus required for this purpose is the same as that used for the reduction of metallic oxides by hydrogen (151). Fill the generating bottle *a* about one-third full of water, and put in some fragments of marble ($CaO,CO_2$); when the apparatus is arranged, pour in from time to time a little hydrochloric acid through the tube *b*, so as to maintain a moderate effervescence (19). Weigh the bulb-tube, and put into it about 10 grains of the mixture of sulphur and charcoal; weigh a second time, to ascertain how much is used in the experiment, and connect the apparatus together. Allow the gas to come over for about five minutes, in order to displace the common air (which might otherwise cause the volatilization of some of the charcoal, by conversion into carbonic acid), and then heat the mixture as long as any sulphur is volatilized. As soon as the apparatus is cold, weigh the bulb-tube again, when the loss of weight will represent the quantity of sulphur contained in the mixture. The percentage of sulphur is then ascertained by calculation.

Weight of mixture : loss of weight : : 100 : percentage of sulphur.

## SECTION III.

### *Preparation of perchloride of iron* ($Fe_2Cl_3$).

158. When metallic iron is heated in a current of chlorine gas, the two substances combine, forming perchloride of iron. The chlorine is generated in a retort *a* (Fig. 61), to the beak of which a tube, bent at right angles *b*, should be adapted by means of a perforated cork. The retort is then charged with 700 grains of a mixture of black oxide of manganese ($MnO_2$) and common salt ($NaCl$), (in the proportion of three parts of the former to four of the latter), on which should be poured two ounces of water. Remove the tube-funnel *b* from the bottle used in the two last experiments (151), and substitute a piece of tubing *e*, sufficiently wide to admit

the bent tube *b*, and reaching nearly to the bottom of the bottle, which should be filled about a fourth part full of water. The rest of the apparatus is the same as that used in the reduction by hydrogen (151), only substituting the straight tube *d*, which may be six or eight inches long, for the bulb-tube before employed; and put into it thirty grains of clean iron wire.

Fig. 61.

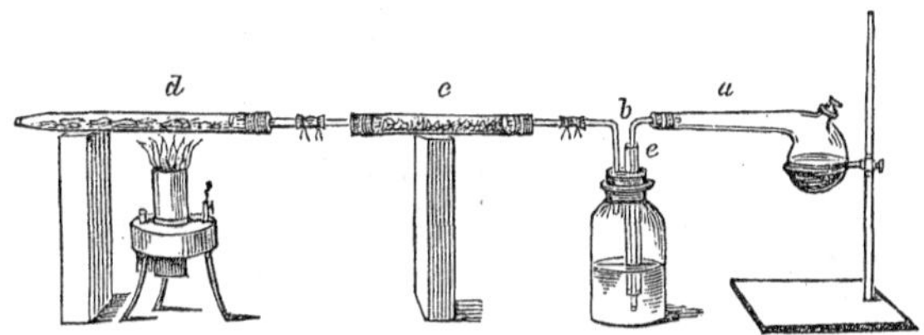


159. When the apparatus is connected together, slowly pour into the retort through a funnel one ounce of strong sulphuric acid ($HO,SO_3$), to disengage the chlorine from the mixture of manganese and salt; and if the gas does not come over properly, apply a very gentle heat.

$$MnO_2+NaCl+2\,(HO,SO_3)=MnO,SO_3+NaO,SO_3+2HO+Cl.$$

The gas, when generated, passes through the water in the bottle *f*, which retains any hydrochloric acid with which it may be impregnated; and having passed over chloride of calcium in the tube *c*, arrives in the tube containing the iron, in a pure and dry state.

When the apparatus is filled with the chlorine, apply a gentle heat to the iron wire, and observe the beautiful scaly crystals of sesquichloride of iron ($Fe_2Cl_3$), which sublime and condense in the cool end of the tube. Remove a few of the crystals from the tube, and remark with what avidity they absorb moisture from the air when exposed to it for a few minutes.

Dissolve a little of the chloride in distilled water, and add ammonia ($NH_3$) in slight excess: the brown precipitate which is produced, is hydrated peroxide of iron (280).

# CHAPTER VIII.

## ALKALIMETRY AND ACIDIMETRY.

### SECTION I.

*Alkalimetry.*

160. The process of alkalimetry has for its object the determination of the quantity of real alkali or alkaline carbonate in any given sample, and is founded on the principle that the quantity of alkali which is neutralized by a known quantity of acid, is always constant and uniform, in obedience to the well-known laws of combination in definite proportions.

Fig. 62.

Alkalimeter Tube.

For example, forty-nine parts, by weight, of oil of vitriol ($HO,SO_3$) combine with thirty-two parts of soda ($NaO$), and when the two substances are brought together in these proportions, the resulting compound (sulphate of soda $NaO,SO_3$) is a perfectly neutral salt: but if the relative quantity of acid or alkali be greater or less than those specified, then there will be an *excess* of one of them present, and the solution containing them will be no longer neutral to test paper. Hence it appears that if we have an unknown quantity of pure alkali in a solution, we can, by treating it with an acid of known strength, and observing how much of the acid is required to neutralize it, readily determine the percentage of potash or of soda in any specimen.

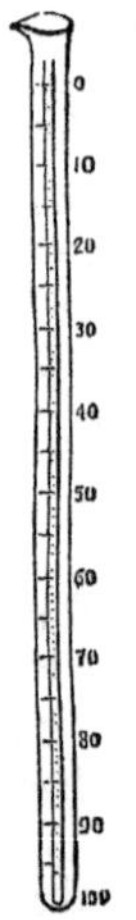

161. The apparatus employed for this purpose is a tube, capable of holding 1000 grains of distilled water, graduated into 100 parts, the divisions being numbered from the open end downwards, as in the figure (Fig. 62). At 65 degrees there is a line scratched, marked *carbonate of potash;* at 54·6, another line, marked *carbonate of soda;* at 49, *potash;* and at 23·5, *soda;* numbers, it will be observed, which, when deducted from 100, bear similar relations to each other as the atomic weights of the

compounds whose names they are associated with. Now, if sulphuric acid of the specific gravity, 1·1268, be poured into the tube up to the point marked by either of these names, the quantity of acid thus measured off will be exactly sufficient to neutralize 100 grains of the alkali specified; and when the tube is filled up to zero with water, it is evident that each division of the tube contains the quantity of acid requisite to neutralize one grain of alkali, being $\frac{1}{100}$th part of the whole. By ascertaining experimentally therefore how many of the divisions full are required to neutralize the alkali present in 100 grains of an impure specimen, that number will represent the percentage of real alkali which it contains.

162. Ascertain the quantity of dry carbonate of soda ($NaO,CO_2$) in a sample of the crystallized salt ($NaO,CO_2+10Aq$). For this purpose weigh out 100 grains of the salt, and dissolve it with the aid of a gentle heat, in about four ounces of water in an evaporating basin. Pour a little of the standard sulphuric acid (specific gravity 1·1268) (which should have been previously prepared and allowed to cool,)[1] into a lipped glass, and thence into the alkalimeter tube, until it reaches the line marked *carbonate of soda*, and fill it with distilled water up to zero. Put two or three small pieces of litmus and turmeric paper into the alkaline solution, which should be kept gently heated over a lamp, and have

Fig. 63.

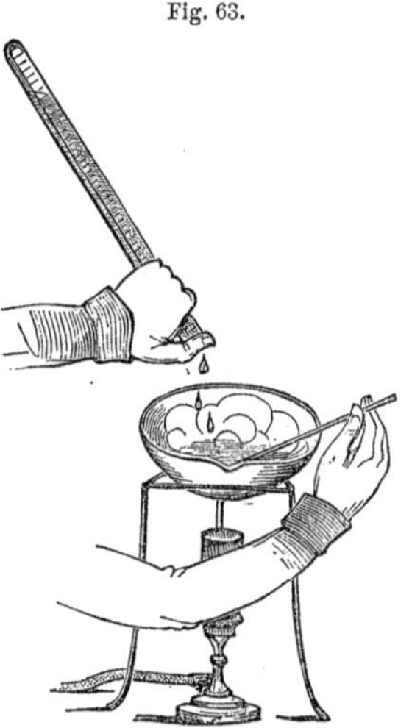

[1] Sulphuric acid of the specific gravity 1·1268, may be prepared by mixing together one part by measure of strong oil of vitriol ($HO,SO_3$) specific gravity 1·84, and eight parts of distilled water. Before taking it into use, it must be tested both as to its specific gravity, and also as to its neutralizing power, which must be ascertained by experiment wit a pure specimen of alkali.

at hand a glass rod to stir it with during the process of neutralization.

163. Having made these preparations, take the tube in the left hand, close the opening tightly with the thumb, and invert the instrument five or six times successively, in order to mix the acid and water thoroughly and uniformly together: then by cautiously relaxing the thumb, allow the acid to fall drop by drop into the alkaline solution, stirring the latter constantly with the glass rod until the litmus begins to turn feebly red. (Fig. 63.) When the change of color begins to appear, wash the sides of the basin by gently agitating the liquid in it, in order to dissolve any of the splashings that may have dried during the process, and escaped the action of the acid. When the point of neutralization is nearly attained, bring one of the pieces of litmus-paper from time to time out of the solution against the heated side of the basin; if the redness disappears, more acid must be added, the reddening being thus proved to have been caused by the carbonic acid dissolved in the water; and the cautious addition of acid must be continued until a permanent feeble red color is obtained.

164. When the neutralization is complete, restore the tube to its vertical position, and remove the thumb (which until now should not have been for a moment removed), scraping it gently, so as to separate most of the adhering acid. Allow the tube to remain upright for a minute or two, in order that the sides may drain, and then observe the degree at which the acid stands, that number representing the percentage of dry carbonate in the sample.[1] The decomposition may be thus expressed:

$$NaO,CO_2+HO,SO_3=NaO,SO_3+HO+CO_2.$$

165. Determine the quantity of soda (NaO) in the same sample. This is done in the way described in the

[1] The atomic weight of crystallized carbonate of soda ($NaO,CO_2+10$ Aq) being 144; and that of the dry salt ($NaO,CO_2$) 54, the percentage of the latter, supposing the crystallized salt to be pure, may be calculated as follows:

144 : 54 :: 100 : $x$ = percentage of dry carbonate of soda.

last experiment, but instead of filling the tube up to the mark *carbonate of soda* with acid, it is filled up to the mark soda, and then up to zero with water.[1]

166. Ascertain experimentally the percentage of potash (KO), and of dry carbonate of potash ($KO,CO_2$), in the crystallized carbonate ($KO,CO_2+2Aq$).[2]

## SECTION II.

### *Acidimetry.*

167. The process of alkalimetry being well understood, that of acidimetry will require but little explanation, as its principle is precisely analogous to that which has been described (160). In the former process the object was to determine the quantity of alkali by the quantity of acid which it was capable of neutralizing; in acidimetry, we have to ascertain the amount of real acid in any solution containing it in an uncombined form. When the acid under examination forms with lime, a salt that is soluble in water, its strength may be ascertained by determining the quantity of marble or carbonate ($CaO,CO_2$) which a given weight of it decomposes and dissolves. This process will serve for nitric, hydrochloric, and acetic acids.

Fig. 64.

Washing bottle.

168. Determine the percentage of nitric acid ($NO_5$) in a specimen of the liquid acid.

Weigh out 150 grains of pounded marble, put it into an evaporating basin, and cover it with about two ounces of distilled water. Pour a little of the acid into a glass, and thence, by means of a dropping-tube (99), transfer exactly 100 grains of it

[1] If the crystallized salt is pure, the percentage of soda may be calculated as follows:

$$\underbrace{\text{Atc. wt. of crystd. carbonate of soda.}}_{144} : \underbrace{\text{Atc. wt. of soda.}}_{32} :: \underbrace{\text{Per cent. of soda.}}_{100 : x}$$

[2] The atomic weight of KO is 48; that of $KO,CO_2$ 70; and that of $KO,CO_2+2Aq$ 88.

into a previously counterpoised capsule. Add this in successive portions to the marble, avoiding too large an addition at once, lest the effervescence should be so violent as to cause some of the liquid to be projected over the sides of the basin and lost.

When the whole of the acid has been added, wash out the dish which contained it, two or three times, with distilled water, and add the washings to the marble; stir the mixture repeatedly with a glass rod, and when the effervescence appears to have nearly ceased, heat it gently over a lamp.

$$CaO,CO_2+\mathit{NO_5}=\mathit{CaO,NO_5}+CO_2.$$

169. While this is going on, fit up a washing bottle (Fig. 64), the tubes for which have been already prepared (94). Two holes must be bored in the cork to fit the tubes, which must be fixed in the manner shown in the figure. Then prepare a filter, according to the directions already given (66); and having moistened it with distilled water, support it over a beaker glass by means of a retort stand or perforated block of wood (Fig. 65). Pour the solution from the evaporating basin down a glass rod into the filter, directing the stream so that it may fall upon the sloping side and not into the apex, lest its force should injure or break through the paper; wash the last portions of marble out of the basin by means of the washing bottle, holding the basin in a nearly vertical position (Fig. 66). When the liquid has for the most part passed through the filter, wash the latter with water from the washing bottle, directing the stream just below the upper edge of the filter, and continuing to wash until a drop of the filtered liquid, when evaporated on a strip of glass or platinum foil, leaves no trace of solid matter. When this is the case, we may be sure that the whole of the

Fig. 65.

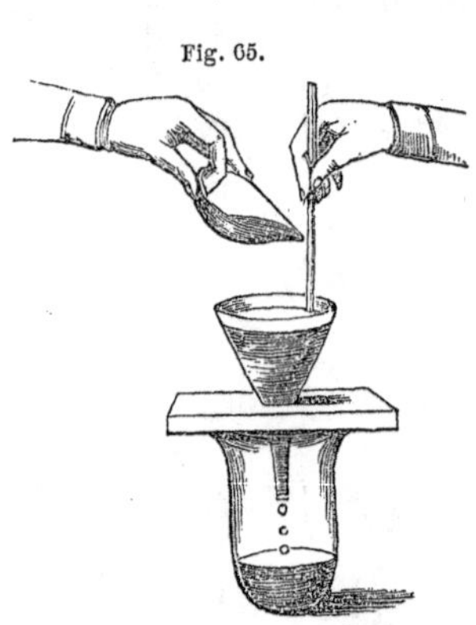

soluble nitrate of lime ($CaO,NO_5$) has been washed out, and that nothing remains but the portion of marble

Fig. 66.

which was not decomposed. This residue must now be thoroughly dried, weighed, and its weight deducted from that of the whole of the marble employed; the difference being of course the weight of that which has been dissolved by the acid. Now since every equivalent of carbonate of lime which has been dissolved, indicates the presence of an equivalent of nitric acid, the quantity of nitric acid in the 100 grains employed may be ascertained by the following proportion:

| Atomic weight of carbonate of lime. | | Atomic weight of nitric acid. | | |
|---|---|---|---|---|
| 50 | : | 54 :: | Quantity of marble dissolved : $x$ = | Percentage of nt. acid. |

170. Ascertain the percentage of hydrochloric acid ($HCl$) in a specimen of the liquid acid. Proceed exactly as in the previous experiment with nitric acid (168), only in the calculation substitute the atomic weight of hydrochloric acid 37, for that of nitric, thus:

50 : 37 : : Quantity of marble dissolved : $x$

171. In the determination of the strength of acids which do not form with lime salts that are soluble in water, the method just described will not, of course, give accurate results; and the following may be adopted, which we will consider as used in the case of sulphuric acid, sulphate of lime being too sparingly soluble to admit of this acid being estimated by the process with marble.

172. Determine the percentage of sulphuric acid ($HO,SO_3$) in a specimen of the dilute acid. Take 100 grains of the acid (weighed in the manner already described (168)) and place it in a tolerably large evaporating basin; dilute it with three or four ounces of water, and wash out with water the dish in which the acid was weighed, so as to avoid the loss of any of the acid: put into the dilute acid two or three pieces of litmus-paper, and heat it gently over a lamp. Dissolve 100 grains of pure crystallized carbonate of soda ($NaO,CO_2+10Aq$) in an ounce and a half of water, applying if necessary a gentle heat, and pour it when cold into the alkalimeter tube (161), washing with water the dish in which the solution was made: then fill the tube up to zero with distilled water. Now take the tube in the left hand, and having closed it securely with the thumb, invert it repeatedly, in order to secure the perfect and uniform mixture of the saline solution and the water. When this has been done, it is evident that each division of the alkalimeter tube must contain in solution one grain of the crystallized carbonate.

The alkaline liquid must now be added gradually to the dilute acid, with the same precautions as in the process of alkalimetry (163), until the red color of the litmus is changed to purplish blue; when this is the case, remove the thumb, and after allowing the tube to stand upright for a minute or two, read off the degree at which the liquid stands; that number representing the number of grains of the carbonate which have been required to neutralize the 100 grains of dilute acid. As each equivalent of the carbonate which is neutralized represents an equivalent of the acid, the following calculation will furnish the percentage of the latter:

| Equiv. of crystallized carbonate of soda. | | Equiv. of sulphuric acid. | | Grains of carbonate used. | | Percentage of acid. |
|---|---|---|---|---|---|---|
| 144 | : | 49 | :: | | : | $x$ |

173. Determine the strength of nitric ($NO_5$) and hydrochloric ($HCl$) acids by this method, and compare the results with those obtained by the decomposition of marble (168, 170).

SECTION III.

*Estimation of Carbonic Acid in Carbonates.*

174. The quantity of carbonic acid ($CO_2$) contained in any carbonate which is readily decomposable by hydrochloric acid, may be determined in the following manner. Take a piece of tube *b* (Fig. 67), five or six inches long, with one end drawn out so as to leave only a small opening, and fill it with fragments of chloride of calcium (734), putting in a loose plug of tow or cotton wool at each end, in the manner already described (152). Bend a piece of quill tubing in the form shown at *d* (88), and by means of perforated corks, connect the two tubes with a flat-bottomed flask, capable of holding ten or twelve ounces of water. Select a tolerably wide test-tube, of such a size as will stand in the flask in an inclined position as shown in the figure, and nearly fill it with strong hydrochloric acid ($HCl$).

Fig. 67.

Carbonic Acid Apparatus.

175. Put into the flask twenty grains of marble ($CaO,CO_2$) in small fragments, and pour upon it about an ounce of water: then cautiously introduce the tube *c* containing the acid, taking care that none of the acid is allowed to come in contact with the marble; connect the chloride of calcium tube with the flask, and accurately weigh the whole apparatus. Now gradually incline the flask, so as to allow the acid to flow slowly upon the marble: the carbonic acid is disengaged with effervescence; is deprived, while passing over the chloride of calcium, of the moisture with which it is impregnated; and passes off through the small aperture in the

tube *b*, leaving the apparatus of course lighter than before.

$$CaO,CO_2+HCl=CaCl+HO+CO_2.$$

176. When the effervescence has ceased, the flask should be gently warmed, and when cool again, the cork may be removed, and air drawn through the flask by means of a small piece of tube, to abstract the whole of the carbonic acid with which it is filled; and which, being heavier than common air (22), would add to its apparent weight. The chloride of calcium tube is then replaced, and the whole apparatus again weighed; the loss of weight being of course that of the carbonic acid expelled. By multiplying this loss by 5 ($20\times5=100$), the percentage of carbonic acid in the marble is obtained.

177. Ascertain in the same way the percentage of carbonic acid in the carbonate ($NaO,CO_2+10Aq$) and bicarbonate of soda ($NaO,HO,2CO_2$).

# PART II.

## ACTION OF REAGENTS ON BASES AND ACIDS.

---

## CHAPTER I.

### SECTION I.

*Introductory.*

178. QUALITATIVE analysis has for its object the determination of the elements of compounds which are contained in any given substance; and those elements and compounds are recognized by certain characteristic appearances which they present when exposed to the action of tests or reagents, or when otherwise treated, as when submitted to heat, &c.

Before proceeding, therefore, to the more complicated processes of analysis, it is advisable that the student should make himself familiar with the action of reagents on the compounds most commonly met with in such investigations, in order to enable him properly to interpret the language in which Nature, through his experiments, replies to his inquiries.

With this purpose in view, he should not merely apply his tests, and superficially note whether a precipitate is or is not formed, but he should endeavor to impress on his recollection the exact appearance which it presents, both as to color and also as to physical structure; whether it is crystalline, curdy, or gelatinous; whether it separates immediately from the solution, or requires time for its development; as well as the action of solvents (as acids and alkalies) upon it. Besides the increased facility which he will thus gain in making subsequent experiments, he will be acquiring habits of close and accurate observation, which will be of infinite value to him, not only in pursuing the study of chemistry, but in almost every occupation of life.

## SECTION II.

### *Classification of Bases and Acids.*

179. In describing the action of reagents, and the rudiments of chemical analysis, all the rarer bases and acids will be omitted, as they would only tend to confuse the student.[1] The following are those which will be treated of, as being most commonly met with in analysis. The bases are classified according to their behavior with *hydrosulphuric acid* (HS), *hydrosulphate of ammonia* ($NH_4S,HS$) and *carbonate of soda* ($NaO,CO_2$). Those in Class IV are precipitated as sulphides from acidified solutions by hydrosulphuric acid; those in Class III are not affected by hydrosulphuric acid when an excess of hydrochloric acid is present, but are thrown down either as sulphides or oxides when their neutral solutions are treated with hydrosulphate of ammonia; those in Class II are not precipitated by either hydrosulphuric acid or hydrosulphate of ammonia, but are thrown down as carbonates, by carbonate of ammonia or of soda: and those in Class I are unaffected by any of those reagents.

BASES.

*Class I.*

| | |
|---|---|
| Potash, . . . . . . . . . . | ($KO$) |
| Soda, . . . . . . . . . . | ($NaO$) |
| Ammonia, . . . . . . . . . . | ($NH_3$) |

*Class II.*

| | |
|---|---|
| Magnesia, . . . . . . . . . . | ($MgO$) |
| Lime, . . . . . . . . . . | ($CaO$) |
| Baryta, . . . . . . . . . . | ($BaO$) |
| Strontia, . . . . . . . . . . | ($SrO$) |

[1] In the Appendix will be found a table showing the behavior of most of the rarer substances with reagents.

*Class III.*

| | |
|---|---|
| Alumina, | ($Al_2O_3$) |
| Oxide of Chromium, | ($Cr_2O_3$) |
| Oxide of Zinc, | ($ZnO$) |
| Protoxide of Manganese, | ($Mn O$) |
| Protoxide of Iron, | ($FeO$) |
| Peroxide of Iron, | ($Fe_2O_3$) |
| Oxide of Nickel, | ($NiO$) |
| Oxide of Cobalt, | ($CoO$) |

*Class IV.*

| | |
|---|---|
| Oxide of Arsenic (arsenious acid), | ($AsO_3$) |
| Arsenic Acid, | ($AsO_5$) |
| Oxide of Antimony, | ($SbO_3$) |
| Protoxide of Mercury, | ($HgO$) |
| Peroxide of Mercury, | ($HgO_2$) |
| Oxide of Lead, | ($PbO$) |
| Oxide of Copper, | ($CuO$) |
| Oxide of Silver, | ($AgO$) |
| Protoxide of Tin, | ($SnO$) |
| Peroxide of Tin, | ($SnO_2$) |
| Oxide of Bismuth, | ($Bi_2O_3$) |

ACIDS.

The acids are divided into two groups, the Inorganic and the Organic.

*Inorganic Acids.*

| | |
|---|---|
| Sulphuric, | ($SO_3$) |
| Phosphoric, | ($PO_5$) |
| Boracic, | ($BO_3$) |
| Carbonic, | ($CO_2$) |
| Silicic, | ($SiO_3$) |
| Hydrochloric, | ($HCl$) |
| Hydriodic, | ($HI$) |
| Hydrosulphuric, | ($HS$) |
| Nitric, | ($NO_5$) |
| Chloric, | ($Cl_5O$) |

*Organic Acids.*

| | |
|---|---|
| Oxalic, | ($HO,C_2O_3$) |
| Tartaric, | ($2\ HO,C_8H_4O_{10}$) |
| Citric, | ($3\ HO,C_{12}H_5O_{11}$) |
| Malic, | ($2\ HO,C_8H_4O_8$) |
| Succinic, | ($HO,C_4H_2O_3$) |
| Benzoic, | ($HO,C_{14}H_5O_3$) |
| Acetic, | ($HO,C_4H_3O_3$) |
| Formic, | ($HO,C_2H_3O_3$) |

180. Should the student find that the action of any test does not agree with that described, it may be owing

to some impurity contained in the test liquid, in which case he may examine it in the manner described in the section on reagents (689).

---

# CHAPTER II.

## METALS BELONGING TO CLASS I.

*Potash, Soda, and Ammonia.*

181. THE three bases belonging to this class are chiefly characterized by the solubility in water of most of their compounds, and the consequent difficulty of obtaining them in an insoluble form, and of separating them from one another in the shape of precipitates. They are distinguished from all other bases by producing no precipitate when tested with either of the three classifying tests—viz., hydrosulphuric acid, hydrosulphate of ammonia, and carbonate of soda, their sulphides and carbonates being all soluble in water.

Solutions of the uncombined or carbonated alkalies are alkaline to test-paper, turning reddened litmus blue, and turmeric brown.

### SECTION I.

*Potash* (KO).[1]

---

A solution of chloride of potassium (*KCl*) may be used.

---

182. When a drop of the solution of a potash salt is evaporated on platinum foil and ignited, it leaves a fixed residue, in which respect it differs from ammonia (192).

183. Observe the action of caustic potash, and carbonate of potash in solution, on *litmus* and *turmeric paper*.

Test the solution in separate test-tubes with *hydrosulphuric acid, hydrosulphate of ammonia*, and *carbonate of soda*. No precipitate is produced in either case.

[1] Those tests which are most characteristic are distinguished by (C).

184. It must be remembered that in many cases, precipitates do not separate *at once* from the solutions, but require *time* for their development. This is especially to be regarded in the precipitation of those salts which are to some extent soluble, as the double chloride of platinum and potassium, bitartrate of potash, ammonio-phosphate of magnesia, and many others. In all such cases, and whenever there is any doubt as to the appearance of a precipitate, it is better to leave it for a time, and not to decide that no precipitation will take place until the mixture has stood twenty-four hours. If, after that period, no precipitate appears, it may be safely inferred that none will afterwards be formed. It is necessary also in these cases, that the solution should be tolerably concentrated.

185. (C) An alcoholic solution of *bichloride of platinum* ($PtCl_2$) when added to neutral or slightly acid potash solutions (especially of chloride of potassium), throws down a fine yellow crystalline precipitate, consisting of the double chloride of platinum and potassium ($KCl,PtCl_2$). The presence of a little free hydrochloric acid assists the formation of the precipitate, especially when the potash-salt is any other than the chloride. If the potash solution is dilute, the precipitate does not form at once; so that it is necessary, in employing this test, when we do not obtain a precipitate immediately, to allow the mixture to stand some time (184) before we decide that no potash is present. In such cases, the best way is to evaporate a mixture of the solution of chloride of potassium and chloride of platinum nearly to dryness on a water-bath (645), and treat the residue with alcohol, which leaves the whole of the double chloride undissolved.

As ammonia produces, with chloride of platinum, a similar precipitate, it is necessary, before deciding that the indication is due to potash, to prove the absence of ammonia (194).

186. (C) Add a solution of *tartaric acid* ($2HO,C_8H_4O_{10}$) in excess to that of the potash salt, which should be either neutral or with a slight excess of alkali. A colorless crystalline precipitate is produced of bitartrate of

potash ($KO,HO,C_8H_4O_{10}$). As in the last test, the precipitate does not appear immediately unless the solution be concentrated; so that it must be allowed to stand a short time before we satisfy ourselves that no potash is present.

The separation of the precipitate, in this and other similar cases, is much assisted by agitating the mixture with a glass rod; wherever the rod has rubbed against the sides of the tube containing it, delicate lines of microscopic crystals are deposited before any precipitate appears in the body of the liquid.

187. (C) Ignite a small fragment of a salt of potash on platinum wire in the deoxidizing flame of the blowpipe (83), and observe the violet color which it communicates to it. A small quantity of the potash (KO) is here deoxidized, and the volatile potassium (K) thus formed, is again oxidized while passing through the outer flame, which combustion is accompanied by the violet flame.

The same color may be observed in the flame of alcohol which contains a little potash in solution.

It is to be observed, that in these experiments the presence of any soda prevents the appearance of the violet tint, on account of the intense yellow color which the latter base gives to the flame (190).

## SECTION II.

### *Soda* (NaO).

---

A solution of sulphate of soda ($NaO,SO_3+10\ Aq$) may be used.

---

188. (C) An alcoholic solution of *bichloride of platinum* ($PtCl_2$) gives no precipitate in solutions of soda salts, even when they are concentrated. If the mixture, however, be allowed to evaporate spontaneously, delicate yellow needle-shaped crystals of the double chloride of sodium and platinum ($NaCl,PtCl_2$) will gradually form, which are so totally different in appearance from the corresponding potash compound (185), besides being

readily soluble in water and alcohol, that the two cannot be mistaken for each other.

189. (C) *Antimoniate of potash* ($KO,SbO_5$) when added to soda salts, either neutral or containing a slight excess of alkali, produces a white crystalline precipitate of antimoniate of soda ($NaO,SbO_5$) (184). If the soda salt under examination contains an excess of acid, it should be neutralized with potash before the addition of the antimoniate, as otherwise a precipitate of antimonic acid ($HO,SbO_5$) or biantimoniate of potash ($KO,2SbO_5$) might be produced, owing to the decomposition of the antimoniate by the free acid.

It is necessary, in employing this test, that both it and the soda solution should be tolerably concentrated, as otherwise no precipitate will be produced (184).

190. (C) When a fragment of a salt of soda is heated before the blowpipe, it communicates an intense yellow color to the flame; the same color is produced, also, when alcohol is mixed with a solution of soda, and burnt.

191. Neither *hydrosulphuric acid*, *hydrosulphate of ammonia*, nor an *alkaline carbonate*, produce any precipitate in solutions of soda, neither does *tartaric acid* (186).

## SECTION III.

*Ammonia* ($NH_3$) *or with one equivalent of water, which all its salts with oxygen acids contain* ($NH_4O$).

---

A solution of muriate of ammonia ($NH_4Cl$) may be used.

---

192. (C) When heated on platinum foil, the salts of ammonia are all decomposed; and (unless the acid, like the phosphoric or boracic, is fixed at a red heat) volatilize completely, leaving, if pure, no fixed residue. They may be in this way readily distinguished from the salts of potash and soda.

193. Like potash and soda, ammonia gives no precipitate with *hydrosulphuric acid*, *hydrosulphate of ammonia*, or an *alkaline carbonate*.

194. (C) *Bichloride of platinum* ($PtCl_2$) throws down

in ammoniacal solutions, which are not very dilute, a yellow crystalline precipitate of the double chloride of platinum and ammonium ($NH_4Cl,PtCl_2$), which is very similar in appearance to that produced in solutions of potash (184, 185).

If we are doubtful whether the precipitate obtained by this test is due exclusively to ammonia, or whether it contains any potash, the precipitated double chloride may be ignited, and the residue digested in water; if the solution thus obtained give any precipitate with nitrate of silver, potash is present. The reason is this: the ammoniacal compound ($NH_4Cl,PtCl_2$) leaves, after ignition, nothing but metallic platinum; while the potash compound ($KCl,PtCl_2$) leaves a mixture of metallic platinum and chloride of potassium, the latter of which, when dissolved in water, and tested with nitrate of silver, gives a precipitate of chloride of silver ($AgCl$) (429).

Fig. 68.

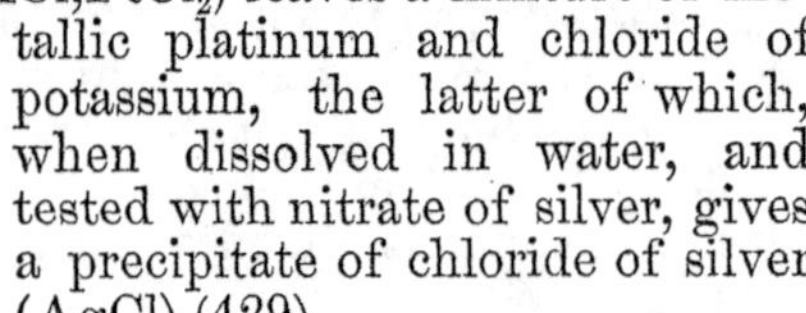

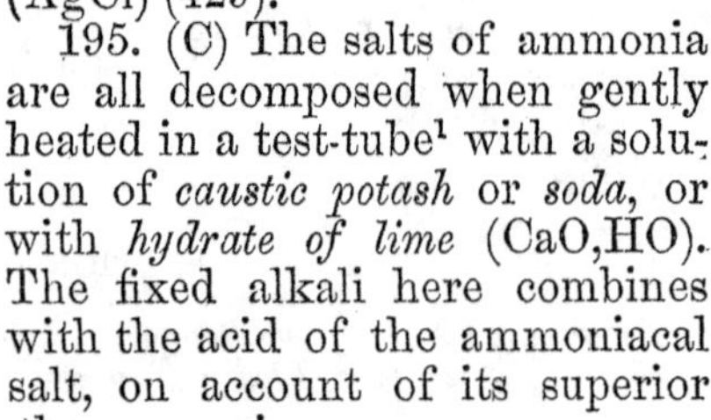

195. (C) The salts of ammonia are all decomposed when gently heated in a test-tube[1] with a solution of *caustic potash* or *soda*, or with *hydrate of lime* ($CaO,HO$). The fixed alkali here combines with the acid of the ammoniacal salt, on account of its superior affinity, and sets free the ammonia.

$$NH_4O,SO_3+KO=KO,SO_3+NH_3+HO.$$

The presence of the free ammonia in the upper part of the tube may be proved,

(*a*) By its well-known odor;

(*b*) By its alkaline reaction on turmeric and reddened litmus-paper, which should be previously moistened, and then held within the tube, care being taken that it does not touch any part of it; and

(*c*) By the production of dense white fumes of muriate of ammonia ($NH_4Cl$), when a rod moistened with dilute

[1] When a liquid is to be boiled in a test-tube, the latter may be conveniently held in a loop of paper or cloth, as shown in Fig. 68.

hydrochloric acid (*HCl*) is held near the mouth of the tube.

196. *Tartaric acid* ($2\ HO, C_8H_4O_{10}$) behaves with ammonia in the same way as with potash, throwing down a colorless crystalline precipitate of bitartrate of ammonia ($NH_4O,HO,C_8H_4O_{10}$), which is, however, rather more soluble than the bitartrate of potash (186).

*Summary of Class I.*

197. From the experiments now described, it appears that the three alkalies may be distinguished from other metallic oxides by their producing no precipitate with either hydrosulphric acid, hydrosulphate of ammonia, or an alkaline carbonate; one or more of which causes, as we shall presently see, a precipitate with all other bases. Hence, if we have a solution which we know to contain some inorganic saline matter, and we find no precipitate produced in it on the application of those tests, we conclude that the base of the salt is either potash, soda, or ammonia.

For the purpose of distinguishing between the three alkalies themselves, we may first test for ammonia, by heating with potash (195). If this is absent, add to a tolerably concentrated solution some bichloride of platinum or tartaric acid (185, 186), which will enable us to distinguish between potash and soda. If these tests give no precipitate, it is probable that the base is soda; which may be confirmed by the behavior of the solution with antimoniate of potash (189), and by allowing the mixture with bichloride of platinum to evaporate spontaneously, when, if yellow needle-like crystals appear, the presence of soda may be considered certain (188).

---

## CHAPTER III.

### METALS BELONGING TO CLASS II.

*Magnesia, Lime, Baryta, and Strontia.*

198. THESE bases are distinguished from the alkalies by the insolubility of many of their salts, especially their

carbonates and phosphates; so that when treated with carbonate or phosphate of soda, they furnish copious precipitates.

SECTION I.

*Magnesia* (MgO).

A solution of the sulphate ($MgO,SO_3+7\,Aq$) is the most convenient for the following experiments.

199. Neither *hydrosulphuric acid* nor *hydrosulphate of ammonia* give any precipitate in solutions of magnesia.[1]

200. (C) Ammonia ($NH_3$) when added to a neutral solution of magnesia, separates a portion of it in the form of hydrate (MgO,HO), which appears as a bulky white precipitate.

$$MgO,SO_3+NH_3+2HO=\mathrm{MgO,HO}+NH_4O,SO_3.$$

The rest of the magnesia remains in solution, in combination with the ammonia and acid, forming a soluble double salt of ammonia and magnesia ($NH_4O,MgO,2SO_3$). Most of these double salts of ammonia and magnesia being soluble in water, and being usually formed when ammoniacal salts are present in excess, the latter have a strong tendency to interfere with the action of the reagents, which in the absence of ammonia produce a precipitate. For example, if the solution of magnesia be mixed with muriate of ammonia ($NH_4Cl$) and then tested with ammonia as above, no precipitate is produced.

201. Collect on a filter, and wash with distilled water, a little of the precipitated magnesia obtained in the last experiments, and place it while moist on yellow turmeric paper; the magnesia being very slightly soluble in water, has an alkaline reaction, and turns it brown.

202. Solution of *caustic potash* ($KO$) precipitates hydrate of magnesia (MgO,HO), especially if the mixture is heated.

$$MgO,SO_3+KO+HO=\mathrm{MgO,HO}+KO,SO_3.$$

[1] When, as is sometimes the case, the hydrosulphate contains free ammonia, it may cause a slight precipitate (200).

Ammoniacal salts (as muriate of ammonia), if present in the solution, prevent the formation of this precipitate, or, if added subsequently, often redissolve it.

203. *Carbonate of potash* ($KO,CO_2$) gives a white precipitate, consisting of basic carbonate of magnesia ($4MgO,3CO_2+4Aq$). A portion of the magnesia remains in solution as bicarbonate, which when boiled is decomposed, and the neutral carbonate ($MgO,CO_2$) being insoluble, is precipitated. Ammoniacal salts, if present, prevent the formation of these precipitates, and redissolve them if subsequently added.

204. *Carbonate of ammonia* ($2NH_4O,3CO_2$) gives no precipitate unless the solution is boiled, and not even then unless it be added sparingly.

205. *Sulphuric acid* ($HO,SO_3$) or *sulphate of soda* ($NaO,SO_3$), produces no precipitate in solutions of magnesian salts, since the sulphate of magnesia is soluble in water.

206. (C) *Phosphate of soda* ($2NaO,HO,PO_5$) gives a white precipitate of phosphate of magnesia ($2MgO,HO,PO_5$) provided the solution is not very dilute, and especially on boiling.

The addition of *ammonia* or its carbonate to the magnesian solution, renders the phosphate of soda a far more delicate test than when used alone, because under those circumstances the double phosphate of ammonia and magnesia ($2MgO,NH_4O,PO_5+12Aq$) is produced, which is less soluble than the phosphate of magnesia, and is consequently thrown down from a more dilute solution than would furnish a precipitate with phosphate of soda alone. If the solution is very dilute, the precipitate does not appear at once, but if allowed to stand some little time, a crystalline deposit of the double phosphate gradually separates (184). Agitation of the liquid with a glass rod hastens the formation of this precipitate; and it is remarkable that if the tube be rubbed at all with the rod during agitation, lines of minute crystals are there first deposited. The same phenomenon occurs in the case of the bitartrate of potash and others, in which the precipitate is slowly deposited from a dilute solution.

As the double phosphate is readily soluble in an excess of acid, and slightly so in water, it is necessary that the solution should be pretty strongly ammoniacal.

It will be observed that in this test, the effect of ammoniacal salts in the solution is the reverse of that before described (200). When mixed with ammoniacal salts indeed, magnesia can be precipitated only by a soluble phosphate.

If the double phosphate be ignited, it is decomposed into phosphate of magnesia ($2MgO,PO_5$), the ammonia and water being expelled.

207. *Oxalate of ammonia* ($NH_4O,C_2O_3$) gives, in tolerably strong solutions, a white precipitate of oxalate of magnesia ($MgO,C_2O_3$), provided no other ammoniacal salts are present.

208. (C) *Baryta water* (*BaO* in water) gradually throws down a white precipitate of hydrate of magnesia ($MgO,HO$) (184). If the sulphate of magnesia be used, the insoluble sulphate of baryta ($BaO,SO_3$) will be thrown down at the same time.

$$MgO,SO_3+BaO,HO=MgO,HO+BaO,SO_3.$$

209. (C) When magnesia or one of its salts is moistened with a solution of nitrate of cobalt ($CoO,NO_5$), and strongly heated before the blowpipe, the mixture assumes a pale flesh or rose color.

## SECTION II.

### *Lime* (CaO).

A solution of chloride of calcium ($CaCl$) or nitrate of lime ($CaO,NO_5+3Aq$) may be used with the liquid tests.

210. Place a small fragment of caustic lime on moistened turmeric paper; the brown color which is produced shows the alkaline nature of lime.

211. *Hydrosulphuric acid* and *hydrosulphate of ammonia* give no precipitate in solutions of salts of lime.

212. *Ammonia* produces no precipitate.

213. *Potash* ($KO$) throws down a white precipitate of hydrate of lime from concentrated solutions, which redissolves when treated with a large quantity of water.

$$CaCl,+KO,HO=\mathrm{CaO,HO}+KCl.$$

If any of the precipitate is insoluble when treated with water, it is probably owing to the potash containing a little carbonate, which would cause the formation of the insoluble carbonate of lime (214). If the solution of hydrate of lime be exposed to the air, it gradually absorbs carbonic acid, and a deposit of carbonate of lime takes place, which dissolves with effervescence in dilute hydrochloric acid.

214. (C) *Carbonate of potash* ($KO,CO_2$) throws down a copious precipitate of carbonate of lime ($CaO,CO_2$), which is readily soluble with effervescence in dilute hydrochloric or nitric acid.

$$CaCl+KO,CO_2=\mathrm{CaO,CO_2}+KCl.$$

The quantity of the precipitate increases on boiling the mixture; and its formation is unaffected by the presence of ammoniacal salts.

215. *Sulphuric acid* ($HO,SO_3$) or *sulphate of soda* ($NaO,SO_3$), when added to concentrated solutions of lime, give an immediate white precipitate of sulphate of lime ($CaO,SO_3+2Aq$).

$$CaCl+NaO,SO_3=\mathrm{CaO,SO_3}+NaCl.$$

If the solution is not concentrated, the precipitate may not appear at once, but will gradually separate in the form of minute crystals (184); and if the solution is very dilute, no precipitation will take place, because the sulphate of lime, being soluble in about 500 times its weight of water, remains dissolved if sufficient water is present. In solutions of sulphate of lime, of course no precipitate is produced by either of these reagents.

216. (C) After having thrown down the sulphate of lime, pour the mixture on a filter, and test the filtered solution with *oxalate of ammonia* (218); sufficient of the sulphate will have been retained in solution to give a very perceptible precipitate with the oxalate. For the success of this experiment, it is necessary that the

liquid is free from any excess of acid, since the oxalate of lime is soluble in most acid solutions (218).

217. *Phosphate of soda* ($2NaO,HO,PO_5$) gives, in neutral or alkaline solutions of lime, a white precipitate of phosphate of lime ($8CaO,3PO_5$), which is readily soluble in dilute hydrochloric acid, and reprecipitated from the acid solution when neutralized with ammonia.[1] The presence of ammonia does not, as in the case of magnesia (206), facilitate the formation of this precipitate.

218. (C) *Oxalate of ammonia* ($NH_4O,C_2O_3$) is an extremely delicate test for lime. When added to a solution containing it even in a highly diluted state, a copious white precipitate of oxalate of lime ($CaO,C_2O_3+2Aq$) is produced, which is one of the most insoluble salts with which we are acquainted.

$$CaCl+NH_4O,C_2O_3=CaO,C_2O_3+NH_4Cl.$$

It is necessary that the solution should contain no excess of acid, as the oxalate of lime is soluble in acid solutions; acetic and oxalic acids, however, do not dissolve it.

219. If alcohol, containing a salt of lime in solution, is burnt, the flame has a reddish tinge, less crimson, however, than that caused by strontia under the same circumstances (236). The salts of lime also communicate a similar color to the blowpipe flame.

## SECTION III.

### *Baryta* (BaO).

A solution of chloride of barium ($BaCl+2Aq$) may be used with the liquid tests.

220. *Hydrosulphuric acid* and *hydrosulphate of ammonia* produce no precipitate with salts of baryta.

221. *Ammonia*, when free from carbonate, gives no precipitate.

222. *Potash* ($KO$) in dilute solutions gives no precipitate; but if the baryta solution be concentrated, it

[1] If the phosphate of soda be added drop by drop to an excess of chloride of calcium, the precipitate consists of ($2CaO,HO,PO_5$).

throws down a bulky crystalline precipitate of hydrate of baryta ($BaO,HO,+9Aq$), which redissolves if water be added.

223. *Carbonate of potash* ($KO,CO_2$), and carbonate of ammonia ($2NH_4O,3CO_2$) throw down a white precipitate of carbonate of baryta ($BaO,CO_2$).

$$BaCl+KO,CO_2=BaO,CO_2+KCl.$$

When sesquicarbonate of ammonia is used, the solution should be mixed with a little free ammonia, and boiled, to decompose any bicarbonate of baryta, which, if present, would remain dissolved. The precipitated carbonate is readily soluble with effervescence in dilute hydrochloric or nitric acid.

224. (C) *Sulphuric acid* ($HO,SO_3$) and *sulphate of soda* ($NaO,SO_3$) produce in solutions of baryta a copious white precipitate of sulphate of baryta ($BaO,SO_3$) even in very dilute solutions.

$$BaCl+NaO,SO_3=BaO,SO_3+NaCl.$$

This precipitate is quite insoluble in hydrochloric and nitric acids, and thus differs from the carbonate formed in the last experiment.

225. (C) Solution of *sulphate of lime* ($CaO,SO_3$) throws down an immediate precipitate of sulphate of baryta ($BaO,SO_3$). This is the most convenient form of applying a very dilute solution of a sulphate (sulphate of lime requiring about 500 times its weight of water to dissolve it), and serves to distinguish baryta from strontia (233).

226. *Phosphate of soda* ($2NaO,HO,PO_5$) causes a white precipitate of phosphate of baryta ($2BaO,HO,PO_5$), which is soluble in free acids, but is reprecipitated when the acid solution is neutralized with ammonia.

$$2BaCl+2NaO,HO,PO_5=2BaO,HO,PO_5+2NaCl.$$

The presence of ammoniacal salts does not affect the formation of this precipitate.

227. *Oxalate of ammonia* ($NH_4O,C_2O_3$) throws down a white crystalline precipitate of oxalate of baryta ($BaO,C_2O_3$) if the solution is not very dilute (184). It requires a much stronger solution of baryta than of lime to cause a precipitate with oxalate of ammonia. The

oxalate of baryta, like that of lime, is readily soluble in free acids.

228. The flame of alcohol, containing a baryta salt, has a yellowish color, in which respect it differs from lime and strontia (219, 236).

## SECTION IV.

### *Strontia* (SrO).

A solution of nitrate of strontia ($SrO,NO_5$) may be used.

229. Neither *hydrosulphuric acid* nor *hydrosulphate of ammonia* produce any precipitate in solutions of strontia.

230. *Ammonia* and *potash* behave with solutions of strontia as with those of baryta; from concentrated solutions, potash throws down the white hydrate of strontia (SrO,HO).

231. *Alkaline carbonates* also act as with solutions of baryta (223), carbonate of strontia ($SrO,CO_2$) being produced.

232. (C) *Sulphuric acid* ($HO,SO_3$) and *sulphate of soda* ($NaO,SO_3$) throw down a white precipitate of sulphate of strontia ($SrO,SO_3$) immediately, if the solution is not very dilute, and after standing a short time if it is so; in the latter case, the precipitated sulphate is in the form of minute crystals.

$$SrO,NO_5+NaO,SO_3=SrO,SO_3+NaO,NO_5.$$

233. (C) Solution of *sulphate of lime* ($CaO,SO_3$) gives no immediate precipitate in solutions of strontia, but if allowed to stand, sulphate of strontia gradually separates. Strontia may thus be distinguished from baryta (225).

234. *Phosphate of soda* ($2NaO,HO,PO_5$) behaves with solutions of strontia as with those of baryta (226).

235. *Oxalate of ammonia* ($NH_4O,C_2O_3$) gives a white precipitate of oxalate of strontia, in strong solutions, but not in dilute.

236. The flame of alcohol in which a salt of strontia is dissolved, or which contains some of the aqueous

solution, assumes a beautiful carmine color, especially if the mixture is stirred. The color of this flame should be compared with that produced when the alcohol contains lime (219). When a salt of strontia is heated before the blowpipe, the same carmine color is communicated to the flame.

*Summary of Class II.*

237. Supposing we have in solution a salt of one of the metals belonging to this class—viz. magnesia, lime, baryta, or strontia, we should be able, without any difficulty, by applying a few of the most characteristic tests, to ascertain which individual of the class it is. Thus we should find that a solution of hydrosulphate of ammonia gave no precipitate, and that an alkaline carbonate gave a white one; from which we should infer that the metal belongs to Class II. We might then test it with a solution of sulphate of lime, which would tell us whether baryta or strontia were present (225); if not, add to a very dilute solution a little oxalate of ammonia, which, if the base were lime, would throw it down as oxalate (218). If neither of these tests gave any indication, add phosphate of soda and ammonia, when if the base is magnesia, the double phosphate of ammonia and magnesia is precipitated (206).

Before finally deciding, however, that the base is either of these, it is always necessary to apply other confirmatory tests in addition to those just mentioned (539).

---

## CHAPTER IV.

### METALS BELONGING TO CLASS III.

*Alumina, Oxide of Chromium, Oxide of Zinc, Protoxide of Manganese, Protoxide of Iron, Peroxide of Iron, Oxide of Nickel, and Oxide of Cobalt.*

238. The metals of the third class are distinguished from those of the first and second, in being precipitated

from their neutral solutions by hydrosulphate of ammonia; and from those of the fourth class in being unaffected (with the partial exception of peroxide of iron (278)) when their solutions, containing a slight excess of acid, are treated with hydrosulphuric acid.

## SECTION I.

### *Alumina* ($Al_2O_3$).

A solution of sulphate of alumina ($Al_2O_3,3SO_3+18Aq$) may be used.

239. *Hydrosulphuric acid* gives no precipitate either in a neutral or acid solution of alumina.[1]

240. (C) *Hydrosulphate of ammonia* ($NH_4S,HS$) when added to a neutral solution, gives a white precipitate of hydrate of alumina ($Al_2O_3,3HO$), and hydrosulphuric acid is at the same time liberated.

$$Al_2O_3,3SO_3+3(NH_4S,HS)+6HO=Al_2O_3,3HO+3(NH_4O,SO_3)+6HS.$$

241. (C) *Ammonia* ($NH_3$) throws down a bulky white gelatinous precipitate, which consists chiefly of hydrate of alumina ($Al_2O_3,3HO$) with a small admixture of ammonia, and a basic salt of alumina; which may be said to be insoluble in an excess of ammonia, although with a very large excess, and under peculiar circumstances, a portion of the precipitate occasionally redissolves.

$$Al_2O_3,3SO_3+3NH_3+6HO=Al_2O_3,3HO+3(NH_4O,SO_3).$$

242. (C) *Potash* ($KO$) also gives a precipitate of hydrate of alumina, which, like that caused by ammonia, usually contains a little basic salt: it differs from it, however, in being entirely soluble in an excess of the precipitant. If the solution in potash is mixed with muriate of ammonia ($NH_4Cl$), the alumina is again precipitated.

$$Al_2O_3,3SO_3+3(KO,HO)=Al_2O_3,3HO,+3KO,SO_3.$$

243. *Carbonate of potash* ($KO,CO_2$) and carbonate of

[1] In most cases of qualitative analysis, hydrosulphuric acid may be applied in the state of solution in water (700).

ammonia ($2NH_4O,3CO_2$) give a precipitate of hydrate of alumina, which is insoluble in excess.

$$Al_2O_3,3SO_3+3(KO,CO_2)+3HO=Al_2O_3,3HO+3(KO,SO_3)+3CO_2.$$

244. *Sulphuric acid* and *sulphate of soda* give no precipitate in solutions of alumina.

245. (C) If a salt containing alumina be moistened with a solution of nitrate of cobalt ($CoO,NO_5$), and heated on charcoal before the blowpipe, it assumes a beautiful sky-blue color, which is very characteristic, as no other substance gives so decided a color, though silica acquires under the same circumstances a tint somewhat similar, but much less intense. The blue color is best seen by daylight, after the mass has cooled, as by candlelight it appears violet.

## SECTION II.

### *Oxide of Chromium* ($Cr_2O_3$).[1]

---

A solution of sulphate of chrome ($Cr_2O_3,3SO_3$) may be used.

---

246. *Hydrosulphuric acid* produces no precipitate either in neutral or acid solutions.

247. (C) *Hydrosulphate of ammonia* ($NH_4S,HS$) when added to neutral solutions of oxide of chromium, throws down a dark green precipitate of hydrated oxide of chromium ($Cr_2O_3,3HO$), which is insoluble in excess.

248. (C) *Ammonia* ($NH_3$) also produces the same precipitate ($Cr_2O_3,3HO$), a small portion of which redissolves in an excess of ammonia, forming a pale pinkish solution, but is again precipitated when the mixture is boiled.

249. (C) *Potash* ($KO$) also throws down the hydrated

[1] It is remarkable that several of the compounds, both soluble and insoluble, of oxide of chromium, which are green by daylight, appear of a reddish-purple color when seen by candlelight. This peculiar form of dichroism is seen to great advantage in a solution of the oxalate of chrome, which is green by daylight, but if held between a candle and the eye, appears purplish crimson. What is still more remarkable is, that if a green object, such as a tree or field, be viewed by daylight through the green solution, it appears of a bright reddish-purple color.

oxide, which is soluble in excess, forming a green solution; if the alkaline solution be boiled for a length of time, the hydrated oxide is again precipitated, leaving the liquid colorless.

250. *Carbonate of potash* ($KO,CO_2$) gives a dull green precipitate of subcarbonate of chromium, which redissolves in large excess of the precipitant.

251. (C) Oxide of chromium, when heated before the blowpipe with borax or microcosmic salt, either in the inner or outer flame, fuses into an emerald-green bead.

252. (C) If it is heated with a mixture of *nitrate of potash* ($KO,NO_5$) and *carbonate of soda* ($NaO,CO_2$), a yellow bead of alkaline chromate is formed. Here a portion of the oxygen of the nitric acid combines with the oxide of chromium ($Cr_2O_3$), converting it into chromic acid ($Cr_2O_6$) or rather ($CrO_3$), which combines with the potash or soda, forming an alkaline chromate ($KO,CrO_3$). If the bead be dissolved in water acidulated with a little nitric acid, the solution will give with salts of lead a bright yellow precipitate of chromate of lead ($PbO,CrO_3$) (363).

## SECTION III.

### *Oxide of Zinc* (ZnO).

A solution of sulphate of zinc ($ZnO,SO_3+7Aq$) may be used.

253. *Hydrosulphuric acid* (HS), when added to a neutral solution of zinc, causes the precipitation of a portion of it as sulphide (ZnS). This test, however, for reasons which will afterwards appear (541), is usually applied to solutions containing a slight excess of hydrochloric or some other acid. For this purpose, acidify a little of the solution in a test-tube with a drop or two of hydrochloric acid (*HCl*) and then test it with hydrosulphuric acid; it will in this case produce no precipitate.

254. (C) *Hydrosulphate of ammonia* ($NH_4S,HS$) when added to a neutral or alkaline solution of zinc, gives a copious curdy precipitate of sulphide (ZnS), which if the zinc salt be pure, is white; but if, as is frequently the

case, any iron is present, the precipitate will be more or less colored, owing to the admixture of a little of the black sulphide of iron (FeS).

$$ZnO,SO_3+NH_4S,HS,=ZnS+NH_4O,SO_3+HS.$$

255. (C) *Ammonia* ($NH_3$) throws down a white gelatinous precipitate of hydrated oxide of zinc (ZnO,HO), which is readily soluble in excess.

$$ZnO,SO_3+NH_3+2HO=ZnO,HO+NH_4O,SO_3.$$

If the ammoniacal solution of the oxide be treated with hydrosulphuric acid, the white sulphide (ZnS) is thrown down.

256. (C) *Potash* (*KO*) behaves in the same manner as ammonia, giving a precipitate (ZnO,HO), soluble in excess. Hydrosulphuric acid throws down the white sulphide from the potash solution.

257. *Carbonate of potash* ($KO,CO_2$) gives a white precipitate of basic carbonate of zinc ($3(ZnO,HO)+2(ZnO,CO_2)$), which is insoluble in an excess of the carbonate. Ammoniacal salts in solution prevent the formation of this precipitate, since they combine with the oxide of zinc, forming double salts, which are soluble in water.

258. *Carbonate of ammonia* ($2NH_4O,3CO_2$) in small quantity throws down the basic carbonate of zinc ($3(ZnO,HO)+2(ZnO,CO_2)$), which readily redissolves in an excess of the ammoniacal salt.

259. *Sulphuric acid*, and *sulphate of soda*, give no precipitate in salts of zinc, because the sulphate of zinc is soluble in water.

260. (C) When oxide of zinc or any of its salts are mixed with carbonate of soda, and heated on charcoal in the inner flame of the blowpipe, the zinc is reduced to the metallic state, in which condition it is volatilized by the heat, and reoxidized while passing through the outer flame; the oxide thus produced is in part deposited on the charcoal in the form of a pale yellow incrustation, which on cooling becomes white (116).

261. (C) If oxide of zinc or its salts be moistened with a solution of nitrate of cobalt ($CoO,NO_5$) and heated in the outer blowpipe flame, the mixture assumes a pale green color, which is very characteristic. Zinc can in

this way be readily distinguished from other substances, especially from alumina and magnesia (245, 209).

### SECTION IV.

### *Protoxide of Manganese* (MnO).

A solution of sulphate of manganese ($MnO,SO_3+7Aq$) may be used.

262. *Hydrosulphuric acid* (HS), when added to an acidified solution (formed by adding a few drops of hydrochloric acid to a little of the solution in a test-tube), gives no precipitate. If the solution is neutral, a partial precipitation of sulphide (MnS) takes place.

263. (C) *Hydrosulphate of ammonia* ($NH_4S,HS$) gives in neutral solutions a flesh-colored gelatinous precipitate of sulphide of manganese (MnS), which is insoluble in excess. If this precipitate be exposed to the air, it is gradually decomposed, and is converted into the dark brown hydrated sesquioxide of ($Mn_2O_3,2HO$), in consequence of the strong affinity of manganese for oxygen, which it absorbs from the air.

264. (C) *Ammonia* ($NH_3$) throws down a white or pale flesh-colored precipitate of hydrated protoxide of manganese (MnO,HO), which if exposed to the air becomes brown, owing to the formation of the sesquioxide ($Mn_2O_3,2HO$), as in the last experiment. $2(MnO,HO)+O=Mn_2O_3,2HO$.

If muriate of ammonia ($NH_4Cl$) is present in the solution, it prevents the precipitation of the hydrated protoxide; or, if added subsequently, redissolves it, owing to the formation of double salts of ammonia and manganese which are soluble in water. If the ammoniacal solution be exposed to the air, the brown sesquioxide is gradually precipitated.

265. *Potash* ($KO$) behaves as ammonia in solutions of manganese: the presence of muriate of ammonia, however, does not altogether prevent the precipitation of the protoxide.

266. *Carbonate of potash* ($KO,CO_2$) or *of ammonia* ($2NH_4O,3CO_2$) throws down a white precipitate of car-

bonate of manganese ($MnO,CO_2$), which is less prone to blacken on exposure than the hydrated oxide.

$$MnO,SO_3+KO,CO_2=MnO,CO_2+KO,SO_3.$$

267. (C) When compounds of manganese are mixed with carbonate of soda ($NaO,CO_2$) and heated on platinum wire in the outer flame of the blowpipe, the manganese becomes more highly oxidized and is changed into manganic acid ($MnO_3$); this combines with the soda to form manganate of soda ($NaO,MnO_3$), which has a characteristic green color. The change is produced still more rapidly if a little nitrate of potash ($KO,NO_5$) be added to the mixture.

268. (C) When mixed with borax ($NaO,2BO_3+10Aq$) or microcosmic salt ($NaO,NH_4O,HO,PO_5+8Aq$), and heated in the outer flame of the blowpipe, the salts of manganese form beads of an amethyst purple color, which is due to the formation of the red oxide ($Mn_3O_4$). If the mixture be heated in the inner flame, the color disappears, owing to the reconversion of the red oxide into protoxide ($MnO$); this loss of color takes place most readily with microcosmic salt.

## SECTION V.

### *Protoxide of Iron* ($FeO$).

A solution of protosulphate of iron ($FeO,SO_3+7Aq$) may be used.

269. On account of the strong tendency of the protoxide of iron to absorb oxygen on exposure to the air, and become sesquioxide, especially in aqueous solutions of its salts, it is difficult to retain the protosalts in solution without some admixture of sesquioxide; so that in testing them, the indications of some of the reagents are frequently more or less different from those caused by a pure protosalt. If the solution of a protosalt be boiled with nitric acid, the protoxide is wholly converted into peroxide. $6FeO+NO_5=3Fe_2O_3+NO_2$.

270. *Hydrosulphuric acid* ($HS$) produces no precipitate in acidified solutions of protoxide of iron: a slight pre-

cipitation of sulphide (FeS) takes place, however, in neutral solutions of some of its salts, especially when the acid with which it is in combination is a feeble one.

271. *Hydrosulphate of ammonia* ($NH_4S,HS$) when added to neutral solutions of protoxide of iron, throws down a black precipitate of sulphide (FeS), which is insoluble in excess.

$$FeO,SO_3+NH_4S,HS+HO=\mathrm{FeS}+NH_4O,SO_3+\mathrm{HS}.$$

272. (C) *Ammonia* ($NH_3$) gives a precipitate of hydrated protoxide of iron (FeO,HO), which is at first nearly white, but almost immediately becomes greenish. If this precipitate be exposed to the air, it absorbs oxygen, and is changed into hydrated sesquioxide or peroxide ($Fe_2O_3$,3HO) which has a reddish brown or rust color. Muriate and some other salts of ammonia, prevent the precipitation of the protoxide by ammonia, forming a solution of a double salt of ammonia and iron, from which the hydrated peroxide is gradually precipitated if exposed to the air.

273. *Potash* (*KO*) behaves as ammonia.

274. *Carbonate of potash* ($KO,CO_2$) produces a precipitate of carbonate of iron ($FeO,CO_2$) which is similar in appearance to the hydrated protoxide (272).

275. (C) *Ferrocyanide of potassium* ($K_2FeCy_3+3Aq$)[1] throws down in solutions of protoxide of iron, a precipitate ($KFe_32FeCy_3$) which is at first almost white, but rapidly changes to pale blue; the color becomes darker on exposure to the air, owing to the absorption of oxygen, which combines with the potassium and a portion of the iron, forming at the same time Prussian blue ($Fe_43FeCy_3$) (282). This change takes place almost immediately if a little nitric acid or chloride of lime be added to the mixture.

276. (C) *Ferridcyanide of potassium* ($K_3,Fe_2Cy_6$)[2] produces in solutions of the protosalts of iron, a beautiful

[1] Ferrocyanogen ($FeCy_3$) which is here combined with potassium, is a hypothetical radical composed of iron in a peculiar state of combination with cyanogen ($C_2N$).—See Fownes' Manual of Chemistry, p. 517.

[2] Ibid. p. 521.

dark blue precipitate, similar in appearance to Prussian blue, consisting of ferridcyanide of iron ($Fe_3,Fe_2Cy_6$).

$$K_3,Fe_2Cy_6+3(FeO,SO_3)=Fe_3,Fe_2Cy_6+3(KO,SO_3).$$

277. When heated with borax before the blowpipe, salts of iron form beads which in the oxidizing flame become orange, and in the reducing flame green; the color being due to the iron in a higher or lower state of oxidation.

### SECTION VI.

*Peroxide or Sesquioxide of Iron* ($Fe_2O_3$).

A solution of the perchloride of iron ($Fe_2Cl_3$) may be used.

278. *Hydrosulphuric acid* (HS) causes in neutral or acidified solutions of the persalts of iron, a slight precipitation of sulphur, which gives the solution a milky appearance. This is owing to the decomposition of hydrosulphuric acid by the peroxide of iron, the hydrogen combining with a portion of its oxygen, reducing it to the state of protoxide, while the liberated sulphur is precipitated in a finely divided state.

$$Fe_2O_3,3SO_3+HS=2(FeO,SO_3)+HO,SO_3+S.$$

279. *Hydrosulphate of ammonia* ($NH_4S,HS$) separates the whole of the iron from solutions of its persalts, as the black sulphide (FeS), the same compound as that produced in the protosalts. The peroxide ($Fe_2O_3$) is in fact first converted into the protoxide ($FeO$) by the deoxidizing affinity of the hydrogen and sulphur in the hydrosulphate, so that the subsequent change is the same as that produced in the protosulphate (271). If the solution of iron is very dilute, no precipitate appears at first, but the solution becomes green, and if allowed to stand a considerable time, the sulphide gradually separates.

280. (C) *Ammonia* ($NH_3$) throws down the hydrated peroxide of iron ($Fe_2O_3,3HO$), in the form of a bulky reddish-brown precipitate, which is insoluble in an ex-

cess of ammonia, and is unaffected by the presence of ammoniacal salts.

281. *Potash* ($KO$) produces the same precipitate ($Fe_2O_3,3HO$), which is insoluble in excess.

282. (C) *Ferrocyanide of potassium* ($K_2FeCy_3+3Aq$) produces in solutions of the persalts of iron a beautiful precipitate of sesquiferrocyanide of iron, or Prussian blue ($Fe_43FeCy_3$).

$$3(K_2,FeCy_3)+2Fe_2Cl_3=Fe_43FeCy_3+6KCl.$$

This is an extremely delicate and characteristic test for the persalts of iron, as the precipitate is produced even in very dilute solutions. In testing for iron with ferrocyanide of potassium, however, it must be borne in mind that when added to a solution containing much free acid, it is partially decomposed, and a little Prussian blue is formed, even when no iron is present. As the presence of free alkalies also interferes with the formation of the blue precipitate, solutions to be tested with it should be as nearly neutral as possible.

283. *Ferridcyanide of potassium* ($K_3,Fe_2Cy_6$) produces no precipitate with persalts of iron; it gives, however, a deep green color to the solution.

284. When heated before the blowpipe, the persalts of iron exhibit the same appearances as those of the protoxide (277), on account of the facility with which the two oxides become converted into one another, according as they are placed in the oxidizing or the reducing flame.

## SECTION VII.

### *Oxide of Nickel* (NiO).

---

A solution of the sulphate of nickel ($NiO,SO_3+7Aq$) may be used.

---

285. *Hydrosulphuric acid* ($HS$) causes no precipitate in acidified solutions of nickel; but if the solution is neutral, especially if the acid of the salt be a feeble one, a partial precipitation of sulphide of nickel ($NiS$) takes place.

286. *Hydrosulphate of ammonia* ($NH_4S,HS$) throws

down, from neutral solutions a black precipitate of sulphide (NiS) which is very slightly soluble in excess, giving a brownish tint to the solution.

287. (C) *Ammonia* ($NH_3$) causes a pale green precipitate of hydrated protoxide of nickel (NiO,HO) which redissolves when the ammonia is added in excess, owing to the formation of a double salt of ammonia and nickel ($NH_4O,NiO,2SO_3$) which is soluble in water. If potash ($KO$) be added to the ammoniacal solution, it reprecipitates the hydrated protoxide of nickel. The presence of ammoniacal salts in the nickel solution prevents the precipitation by ammonia.

288. *Potash* ($KO$) also throws down the hydrated oxide of nickel (NiO,HO), which is insoluble in an excess of potash.

289. *Carbonate of potash* ($KO,CO_2$) gives a precipitate of carbonate of nickel ($NiO,CO_2$), together with a little hydrated oxide, insoluble in excess.

290. *Carbonate of ammonia* ($2NH_4O,3CO_2$) produces the same precipitate, which redissolves in excess.

291. (C) *Cyanide of potassium* ($KCy$)[1] throws down a precipitate of cyanide of nickel (NiCy), which has a yellowish green color: it redissolves in an excess of the alkaline cyanide, forming a dull yellow solution of the double cyanide of nickel and potassium ($NiCy,KCy$), from which the cyanide of nickel is again precipitated on the addition of dilute sulphuric or hydrochloric acid. If the acid mixture be boiled, the precipitate again dissolves, forming a solution of sulphate or chloride of nickel.

292. (C) When heated with carbonate of soda ($NaO,CO_2$) or borax ($NaO,2BO_3+10Aq$) in the inner flame of the blowpipe, the compounds of nickel are reduced to the metallic state, forming gray-colored beads, owing to the minutely-divided metal being held in suspension by the melted flux: if the latter be dissolved out with water, the precipitated metal will be found to be magnetic. In the outer flame with borax, the color of the

[1] Cyanogen (Cy) which is here combined with potassium, is a compound of carbon and nitrogen ($C_2N$).—See Fownes' Manual of Chemistry, p. 504.

bead is usually violet while hot, becoming brown or yellow on cooling. With microcosmic salt, the bead is reddish while hot, but loses the color more or less entirely when cold.

## SECTION VIII.

### *Oxide of Cobalt* (CoO).

A solution of the nitrate ($CoO,NO_5+6Aq$) or chloride ($CoCl$) may be used.

293. *Hydrosulphuric acid* (HS) gives in acidified solutions no precipitate. If the solution is neutral, a slight precipitation of the black sulphide of cobalt (CoS) takes place.

294. *Hydrosulphate of ammonia* ($NH_4S,HS$) throws down from neutral solutions a copious black precipitate of the sulphide (CoS), which is insoluble in excess, and also in hydrochloric acid.

$$2CoCl+NH_4S,HS=2\mathrm{CoS}+NH_4Cl+HCl.$$

295. (C) *Potash* ($KO$) throws down a precipitate of a blue color, consisting of basic salts of cobalt, which soon becomes greenish if exposed to the air, owing to the absorption of oxygen; and lastly, especially if the solution be boiled, dirty red, owing to the formation of hydrated oxide of cobalt (CoO,HO). The precipitate is insoluble in excess of potash.

296. (C) *Ammonia* ($NH_3$) behaves as potash, but the precipitate readily redissolves in an excess, forming double salts of cobalt and ammonia, which are soluble in water. If the ammoniacal solution is exposed to the air, it gradually becomes darker, owing to the absorption of oxygen, and formation of peroxide of cobalt ($Co_2O_3$).

297. *Carbonate of potash* ($KO,CO_2$) produces a pale pink precipitate, which is a mixture of carbonate of cobalt ($\mathrm{CoO,CO_2}$) and hydrated oxide (CoO,HO).

298. (C) *Cyanide of potassium* ($KCy$) when added to a solution of cobalt, especially when a slight excess of hydrochloric acid is present, gives a pale brown precipitate of cyanide of cobalt (CoCy), which, when heated

with an excess of cyanide of potassium, readily redissolves, forming a soluble double cyanide of cobalt and potassium ($K_3Co_2Cy_6$). The addition of sulphuric acid causes no precipitate in this solution (291).

299. (C) The compounds of cobalt, when fused with borax ($NaO,2BO_3+10Aq$), either in the inner or outer flame of the blowpipe, form beads of an intense blue color, or if there is much cobalt present, nearly black; this appearance is very characteristic. Microcosmic salt acts with cobalt in a similar manner, but in a less marked degree. When mixed with carbonate of soda, and heated on charcoal in the deoxidizing flame, oxide of cobalt is reduced to the metallic state, forming a magnetic powder.

---

# CHAPTER V.

## METALS BELONGING TO CLASS IV.

*Arsenic (of which there are two oxides, both having acid properties—namely, Arsenious Acid and Arsenic Acid), Oxide of Antimony, Protoxide of Mercury, Peroxide of Mercury, Oxide of Lead, Oxide of Copper, Oxide of Silver, Protoxide of Tin, Peroxide of Tin, and Oxide of Bismuth.*

300. THESE metals are distinguished from those of the three preceding classes, by being precipitated from their acidified solutions when treated with hydrosulphuric acid. It is remarkable that nearly all the metals whose compounds are most eminently poisonous belong to this class, and as these are the most important, especially to the medical student, they are placed first, and described in the order of their importance. The oxides of arsenic, though possessing acid properties, and consequently belonging strictly to the chapter on acids, have so many peculiarities in common with this class of oxides, that I have included them in it.

10*

## SECTION I.

### *Arsenious Acid or Oxide of Arsenic* ($AsO_3$).

301. On account of the highly poisonous nature of arsenic, great care should be taken, in the following experiments, not to use more than is absolutely necessary to exhibit its peculiarities. In all these experiments (except those of solution), a fragment the size of a small pin's head is quite sufficient. There is also another advantage in using such small quantities—namely, that in most medico-legal investigations, the quantity to be looked for is equally or even more minute, and it is consequently very important that the student should make himself familiar with the appearances which would, under these circumstances, present themselves.

301 *a*. (C) The following experiment should only be made either in the open air or in a well-ventilated room, on account of the poisonous properties of the arsenic vapor. If a small fragment of arsenious acid be heated on charcoal before the blowpipe, it is wholly volatilized, and a smell of garlic will generally be perceptible, especially when it is subjected to the reducing flame. Both metallic arsenic and its oxide are volatile when heated, but the fumes of the latter have no smell. The odor of arsenic vapor appears to be due to the metal while undergoing oxidation, and may be caused perhaps by the formation of a lower oxide than the arsenious acid; it is always observable when metallic arsenic is volatilized in contact with the air.

302. (C) Place a fragment of arsenious acid in a narrow tube (102), and apply a gentle heat with the blowpipe. It sublimes without decomposition, and condenses in the cool part of the tube, in the form of minute sparkling octohedral crystals (Fig. 69), which should be examined with a lens, as they are highly characteristic. The size and regularity of the crystals depend on the slowness with which the vapor is condensed. If the surface of the glass on which the condensation takes place is quite cold, the sublimate is often amorphous, as may be seen by holding a piece of cold glass in the

fumes given off by a little arsenious acid, heated on charcoal (301). The best way to obtain large and well-defined crystals, is to put a few grains of arsenious acid at the bottom of a common test-tube, and allow it to stand on a tolerably hot sand-bath for half an hour, the lower part only of the tube being embedded in the sand. If a small strip of flat glass be also placed inside the tube, a portion of the acid will condense upon its surface; thus furnishing a convenient specimen for microscopic examination.

Fig. 69.

Crystals of Arsenious Acid.

303. (C) Mix a little oxide of arsenic ($AsO_3$) with black flux (751), which if at all damp should be previously dried on the sand-bath, and heat a little of the mixture in a clean tube before the blowpipe. The arsenious acid is deoxidized by the carbon of the flux, and the metallic arsenic thus reduced sublimes, and condenses in the upper part of the tube, forming a more or less brilliant metallic crust, *a* (Fig. 70). $AsO_3+3C=As+3CO$.

Fig. 70.

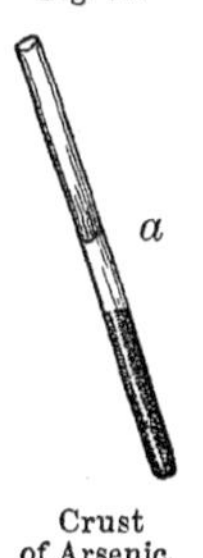


Crust of Arsenic.

If heat be now applied to the sublimate, it will again volatilize, and if any of the vapor escapes from the tube, it may be recognized by its characteristic odor of garlic (301 *a*).

304. (C) Cut off by means of a file, the portion of the tube containing the crust, break it into fragments, and place some of them in another tube. Sublime the arsenic backwards and forwards two or three times in the tube, and observe the gradual conversion of the metal into crystalline arsenious acid, which is formed by the action of the atmospheric oxygen contained in the tube.

305. Boil a few grains of arsenious acid with water (in which it is sparingly soluble), in a flask: filter the solution from the undissolved portion, and retain it for testing.

306. Repeat the last experiment, with the addition of a few drops of solution of potash (*KO*) to the water, and observe the increased solubility of the arsenic, owing to the formation of arsenite of potash ($KO,AsO_3$). Retain the solution for testing.

307. (C) *Hydrosulphuric acid* (HS), when passed through a solution of arsenious acid or of a neutral arsenite (699), causes a slow and gradual precipitation of tersulphide of arsenic, or sulpharsenious acid ($AsS_3$), which it will be observed is analogous in composition to the oxide ($AsO_3$), three equivalents of sulphur being substituted for three equivalents of oxygen. $AsO_3+3HS=AsS_3+3HO$.

If the solution be acidified, however, with a few drops of hydrochloric acid, a much more rapid and complete decomposition takes place; and if the gas be passed through the solution for some time, a complete separation of the arsenic may in this way be effected. The sulphide of arsenic thus formed has a bright light-yellow color; it is insoluble or nearly so in dilute hydrochloric acid, but readily soluble in solutions of the alkalies or their carbonates. Boiling nitric acid ($NO_5$) also dissolves it with decomposition, forming sulphuric and arsenic acids.

$$AsS_3+5NO_5=AsO_5+3SO_3+5NO_2.$$

308. (C) Filter the yellow sulphide formed in the last experiment, and dry a portion of it at a gentle heat on the sand-bath: mix a little of it with black flux (303), heat it in a tube, and observe the formation of a metallic crust of arsenic. $AsS_3+3(KO,CO_2)+2C=As+3KS+4CO_2+CO$.

309. *Hydrosulphate of ammonia* ($NH_4S,HS$), when added to a neutral solution of an arsenite, also causes the formation of the yellow sulphide ($AsS_3$) which however does not precipitate, but remains dissolved as the double sulphide of arsenic and ammonia ($NH_4S,AsS_3$). If an acid be added in excess to the mixture, the sulphide of arsenic is immediately precipitated, of a somewhat lighter color than that thrown down by hydrosulphuric acid, owing to the admixture of a little sulphur derived from the hydrosulphate of ammonia (440).

310. (C) *Nitrate* ($AgO,NO_5$) or *ammonio-nitrate* ($AgO, 2NH_3,NO_5$) *of silver*, throws down in neutral solutions of arsenic, a canary-colored precipitate of arsenite of silver ($2AgO,AsO_3$), which is soluble both in ammonia and nitric acid. It must be remembered that *phosphate of soda* also produces, with nitrate of silver, a similar precipitate, which is equally soluble in nitric acid and ammonia (378).

311. (C) *Sulphate* ($CuO,SO_3$), or *ammonio-sulphate* ($CuO,2NH_3,HO,SO_3$) *of copper*, produces in neutral arsenical solutions, a delicate green precipitate of arsenite of copper ($2CuO,AsO_3$), which dissolves readily both in ammonia and nitric acid, forming a rich blue solution. It must be borne in mind, in employing this test, that a similar precipitate is produced when the solution of copper is added to a liquid containing decoction of onions and some other vegetable substances, though no arsenic may be present.

*Marsh's Test.*

312. (C) It is well known that when zinc is treated with dilute sulphuric acid, it is oxidized at the expense of the oxygen of the water, and hydrogen gas is given off (12). If in addition to the zinc and dilute sulphuric acid, either of the oxides of arsenic are present, the zinc abstracts oxygen from them as well as from the water; and the metallic arsenic thus formed, combines, at the moment of its liberation, with some of the hydrogen simultaneously produced, and forms a gaseous compound called *arseniuretted hydrogen* ($AsH_3$), which passes off mixed with the excess of hydrogen.[1]

$$AsO_3+6Zn+6(HO,SO_3)=6(ZnO,SO_3)+AsH_3+3HO.$$

Now if this arseniuretted hydrogen is heated strongly either by burning in the air, or by passing through a red-hot tube, it is decomposed, and metallic arsenic or its oxide is deposited in the solid state, while the liberated hydrogen passes off.

[1] It must be borne in mind that this gas, like most of the other compounds of arsenic, is highly poisonous; so that the experiment should never be performed in a close room, but in the open air or in a well-ventilated apartment.

313. Several forms of apparatus have been contrived for making use of this property in the detection of arsenic;—of these the following is in practice the most convenient: The bottle *a* (Fig. 71) should be capable of containing six or eight ounces of water, and is connected by means of a perforated cork with the tubes *b* and *c*, which should be about half an inch in diameter: to the latter is attached by means of a cork, the tube *d*, which should be made of hard German glass, bent at a right angle, having the end *e* drawn off so as to diminish the aperture. A few fragments of zinc are placed in the bottle, and when the cork with its tubes is attached, pour a little dilute sulphuric acid down the tube *b*, which should reach nearly to the bottom of the bottle, and allow the gas (hydrogen) to be given off for five minutes.[1] Then heat the narrow tube with a spirit lamp at the point *d*, and observe carefully whether there is any deposit produced inside the tube: this precaution is necessary, since some kinds of sulphuric acid, and also of zinc, contain traces of arsenic. If no stain is produced, it may be assumed that the materials are pure.

Fig. 71.

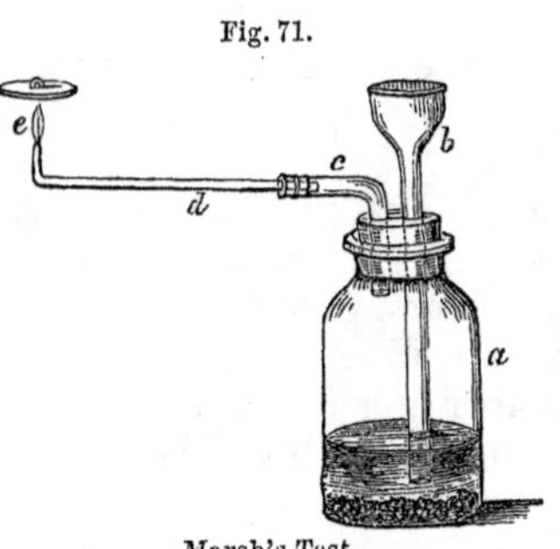


Marsh's Test.

314. The solution containing (or suspected to contain) arsenic, acidified with a few drops of hydrochloric acid, is now introduced through the tube *b*, the heat being still applied to the narrow tube at *d*, as before. If arsenic is present in the liquid, it will cause the formation of arseniuretted hydrogen; which on passing through the heated tube is decomposed, and the arsenic deposited, not exactly at the heated point, but a quarter

[1] The reason why it is not safe to apply the heat at once, is, that a mixture of hydrogen and common air is highly explosive (17), so that it is necessary to allow time for the whole of the common air to be expelled by the hydrogen; as otherwise serious injury might be caused by an explosion of the mixed gases.

or half an inch beyond, in consequence of its volatility. The metallic crust thus formed may be volatilized backwards and forwards in the tube, by heating it with the flame of a spirit lamp.

315. (C) The arsenical crust may also be obtained in another way—namely, by lighting the jet of gas as it issues from the aperture *e*, and holding in the flame a small porcelain plate (for which purpose the lid of a porcelain crucible answers extremely well), when the metallic arsenic will be deposited in the form of a dark shining spot: if the porcelain plate be raised a little, so as to be out of the flame, the arsenic in the state of vapor becomes oxidized while passing through the air, and a white deposit of arsenious acid is formed on the plate. By applying heat to the dark spots, they are readily volatilized, and the fumes will be found to have the characteristic odor of garlic.

A few of these spots should be retained for further examination, and for comparison with those of antimony (320).

316. Hold a short test-tube with the mouth downwards, just above the apex of the flame (Fig. 72) for a few moments, so as to collect some of the arsenious acid formed by the oxidation of the arsenic vapor, and reserve the tube for comparison with antimony (323).

Fig. 72.

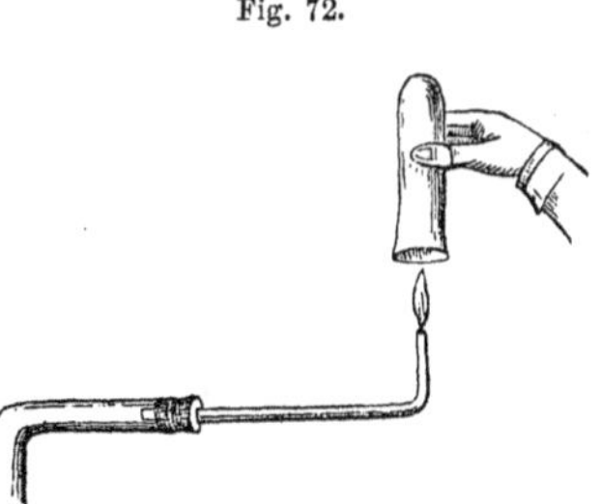

317. Marsh's test as just described is so extremely delicate, that it is capable of detecting arsenic in a solution containing the millionth of its weight of the acid, and may be considered the most conclusive test which we possess. It is however liable to this objection, which is, in practice, easily overcome. It is found that antimony, when present in a mixture of zinc and dilute sulphuric acid, combines with the liberated hydrogen, precisely in the same way as arsenic, forming an analogous compound called antimoniuretted hydrogen ($SbH_3$); which when, heated, is

decomposed, and the metallic antimony is at the same time deposited. Hence it is extremely important that we should be able to distinguish accurately between them, as otherwise we should not be sure whether the crust produced by Marsh's test were due to arsenic or antimony. One or two experiments are generally sufficient to enable us to do this.

318. For the purpose of comparison, empty the zinc and sulphuric acid from the bottle used for the arsenic experiments (313), and substitute fresh zinc and acid. When the gas has been coming off about five minutes (note to 313), pour in a few drops of a solution of the double tartrate of antimony and potash ($KO,SbO_3,C_8H_4O_{10}+2Aq$) and apply heat as before at the point *d*. A crust of antimony will be deposited *at the heated point*, and not, as in the case of arsenic, at a little distance from it; this is owing to the antimony being less volatile than arsenic, and it will be found impossible to volatilize it by the heat of a common spirit lamp. In this respect, therefore, we are enabled in some measure to judge whether the stain is due to antimony or arsenic.

319. Light the jet of gas that issues from the aperture *e* as in (315), and hold over the flame a porcelain plate as before: a deposit of metallic antimony will be formed similar to that of arsenic, but blacker and less shining.

Prepare a few of these spots for comparison with those of arsenic formed in (315).

320. (C) Apply the heat of a spirit lamp to one of each kind of spot, and observe the superior volatility of the arsenic, and the garlic odor of its vapor.

321. (C) Moisten one of each kind of spot with hydrosulphate of ammonia ($NH_4S,HS$) which for this purpose should contain an excess of sulphur (710), and observe that the antimony is *immediately dissolved*, while the arsenic remains nearly unaffected for a considerable length of time. This is a most valuable means of distinguishing between them, and was first observed by Dr. Guy.

322. (C) If the spots be moistened with a solution of

chloride of lime (*CaOCl*), the arsenic will dissolve, while the antimony will remain unaffected.

323. (C) The following may also be taken as a distinguishing test between the arsenic and antimony when Marsh's process is followed. Light the jet of gas issuing from the apparatus, and hold over it a short tube as in (316), so as to collect a little of the oxide of antimony ($SbO_3$) formed by the oxidation of the antimonial vapor. Compare the sublimate thus formed, with that of arsenious acid, and observe the more crystalline appearance of the latter. When the tube is cold, pour in a little water, and treat the arsenious acid in the same way: observe the latter dissolves in the water, while the oxide of antimony remains insoluble. The solution of arsenious acid may then be divided into three portions and tested; the first with *hydrosulphuric acid* (307); the second with *ammonio-nitrate of silver* (310); and the third with *ammonio-sulphate of copper* (311).

These experiments, in conjunction with the other liquid tests, will be found sufficient to prevent the possibility of error in the use of Marsh's test.

### *Reinsch's Test.*

324. (C) This test is founded on the circumstance that when a metal, such as copper, is heated in a solution of another metal more electro-negative than itself, the latter is separated in the metallic state, and deposited on the surface of the former, which is at the same time dissolved in atomic proportion. A little of the solution containing arsenic is acidified with a few drops of hydrochloric acid, and boiled in a test-tube with a strip or two of clean copper foil: the arsenic, being more electro-negative than the copper, is deposited on the surface of the foil, and the whole is in this way separated from the solution.

$$AsO_3 + 3Cu = As + 3CuO.$$

325. The appearance of a metallic deposit on the copper is not, however, necessarily a proof of the presence of arsenic, since other metals (as bismuth, silver, mercury, or antimony) would produce a similar incrustation, being all more electro-negative than copper. Arsenic,

11

however, is readily distinguished from any of these in the following manner.

Take the copper strips out of the solution, and dry them cautiously between folds of filtering paper, or with a very gentle heat: place them in a clean dry test-tube, and apply heat, when the arsenic will be volatilized, and, becoming oxidized by the air contained in the tube, will form a crystalline sublimate in the upper part (302).

Had the deposit on the copper been either of the other metals (with the possible exception of antimony and mercury), it would not have been volatilized when heated: if it were mercury, minute globules of the metal would have condensed in the cool part of the tube: and had it been antimony, a higher degree of heat would have been necessary to sublime it;—the sublimate would have been white and amorphous instead of crystalline;—and when treated with water, would prove insoluble, while the arsenious acid would dissolve, and the solution, on being tested, would show the presence of arsenic.[1]

This excellent test may be considered almost equal to Marsh's both in point of delicacy and freedom from sources of error.

*Arsenic Acid.*

326. Mix a little arsenious acid with nitre ($KO,NO_5$) and heat it in a tube. The nitric acid of the nitre gives up a portion of its oxygen to the arsenic, forming arsenic acid ($AsO_5$).

$$2AsO_3+2NO_5=2AsO_5+NO_2+NO_4.$$

Dissolve the fused mass in water, neutralize the solution with dilute nitric acid, and test it with *nitrate of silver* ($AgO,NO_5$): a reddish-brown precipitate of arseniate of silver ($3AgO,AsO_5$) is thrown down, which is soluble in nitric acid, and also in ammonia.

[1] As the hydrochloric acid of commerce frequently contains traces of arsenic, it is always absolutely necessary, in medico-legal investigations, to ascertain whether the acid employed is perfectly free from it; this is easily done by boiling a little of the acid, diluted with distilled water, in a test-tube with copper foil, which should then be dried and heated in a clean tube, when if arsenic is present it will sublime.

*Detection of Arsenic in Organic Mixtures.*

327. In most cases of medico-legal investigation as to the presence of arsenic, we have to deal with mixtures containing a considerable quantity of organic matter both liquid and solid, which seriously interferes with the action of the tests. Several methods have been employed to get rid of these matters, but the following is perhaps the simplest, and at the same time the most effectual: it is a modification of Reinsch's test.

If the organic mixture suspected to contain arsenic is fluid, it is, previous to filtration, boiled for half an hour with about one-tenth of its bulk of strong hydrochloric acid, the purity of which should of course be ascertained (see note to 325); and if necessary, filtered from any solid matter. It is then boiled with copper foil, when the arsenic, if present, is deposited on the copper, which must be subsequently heated in a tube according to the directions already given (325).

If the matter to be examined is solid, it is treated with dilute hydrochloric acid containing about one-tenth of the strong acid, boiled for half an hour or an hour, filtered if necessary, and then boiled with copper as before.

If the arsenic is present only in very small quantity, a quarter of an hour may elapse before the deposition takes place; and if it does not then appear, the boiling should be continued half an hour or even longer, before we finally conclude that no arsenic is present.

For further particulars on this subject, the student may refer to Dr. Christison's Treatise on Poisons, or to Dr. Guy's excellent work on Forensic Medicine.

## SECTION II.

### *Oxide of Antimony* ($SbO_3$).

For the liquid tests, a solution of the double tartrate of antimony and potash ($KO,SbO_3,C_8H_4O_{10}+2Aq$), or of chloride of antimony ($SbCl_3$) in hydrochloric acid, may be used.

328. (C) Heat a small crystal of the double tartrate in a tube, and observe that it decrepitates and blackens, owing to the decomposition of the vegetable acid ($C_8 H_4O_{10}$) and the consequent deposition of charcoal. Ignite the residue, which consists of charcoal, carbonate of potash, and oxide of antimony, on charcoal in the deoxidizing flame of the blowpipe, when the oxide of antimony will be reduced, and small globules of the metal will appear: a portion of the reduced metal volatilizes with the heat, becomes reoxidized while passing through the outer flame, and the oxide thus produced is deposited on the charcoal, either in the form of a white powder or in crystalline needles.

329. (C) When a stream of *hydrosulphuric acid* (HS) is passed through a solution of antimony acidified with a little hydrochloric acid, an orange-red precipitate of sulphide of antimony ($SbS_3$) is produced, which is soluble in alkaline solution, and difficultly so in hot hydrochloric acid. If the solution is neutral, the precipitation takes place but imperfectly, and in alkaline solutions not at all.

330. *Hydrosulphate of ammonia* ($NH_4S,HS$) when added in a small quantity, gives an orange precipitate of sulphide ($SbS_3$), which redissolves in an excess of the hydrosulphate. If the alkaline solution thus formed be neutralized with an acid, the sulphide is reprecipitated, mixed with a little sulphur (309).

331. *Ammonia* ($NH_3$), *Potash* ($KO$), or their carbonates, throw down from solutions of chloride of antimony ($SbCl_3$), but not in solutions of the double tartrate, a white precipitate of oxide of antimony ($SbO_3$), which is soluble in excess of potash, but insoluble or nearly so in the other solutions.

332. (C) If a solution of chloride of antimony in hydrochloric acid be diluted with a good deal of *water*, a white precipitate of basic oxichloride of antimony ($Sb Cl_3,5SbO_3$) is produced, which if allowed to stand for some time, becomes crystalline.

A similar precipitate is formed under the same circumstances in solutions of bismuth (394): the bismuth precipitate may be distinguished by its insolubility in tartaric acid ($2HO,C_8H_4O_{10}$), in which the oxichloride of antimony is soluble.

333. A piece of clean *zinc* or *copper* causes a precipitation of antimony in the metallic state (324).

334. (C) When oxide of antimony is present in a mixture of zinc and dilute sulphuric acid, the antimony is reduced and combines with the hydrogen, as already described in the case of arsenic (312), forming antimoniuretted hydrogen ($SbH_3$), which is decomposed when burnt, or when passed through a heated tube, with the formation of a deposit of metallic antimony. This experiment has already been described (318).

For the methods of distinguishing between antimony and arsenic, see (317) to (323).

## SECTION III.

### *Protoxide of Mercury* (HgO).

For the first five experiments, calomel (HgCl) may be used; for the rest, a solution of the protonitrate ($HgO,NO_5$) may be taken.

335. (C) Heat a small fragment of calomel (not larger than a small pin's head) in a clean tube. It becomes pale yellow, and, being volatile, it sublimes and condenses in the upper part of the tube; on cooling, the color disappears.

Fig. 73.

336. (C) Dry a small piece of carbonate of soda ($NaO,CO_2$) either in a tube or on a piece of charcoal; mix with it a little calomel, and put the mixture into a tube, *a*, (Fig. 73); then cover it with a layer of carbonate of soda in powder, about a quarter of an inch deep, *b*, and apply heat. The calomel is decomposed, and minute globules of metallic mercury condense in the cool part of the tube at *c*.

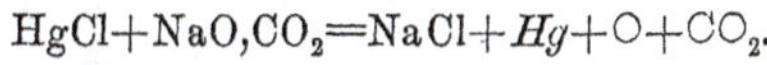

$$HgCl+NaO,CO_2=NaCl+Hg+O+CO_2.$$

337. Boil a little calomel with distilled water in a test-tube; pour off the water into another tube, and test it with hydrosulphate of ammonia (340); no effect is

produced, proving that the calomel is insoluble in water.

338. (C) *Potash* ($KO$) or *ammonia* ($NH_3$) poured on the calomel, decomposes it, turning it black, owing to the formation of the protoxide (HgO). Chloride of potassium ($KCl$) is at the same time formed.

$$HgCl+KO=HgO+KCl.$$

339. (C) Boil a little calomel in fine powder with a solution of *protochloride of tin* ($SnCl$): after some little time the mercury is reduced to the metallic state, owing to the strong affinity which the protochloride of tin has for an additional equivalent of chlorine, which converts it into the bichloride ($SnCl_2$). $HgCl+SnCl=Hg+SnCl_2$.

340. (C) When a solution, either neutral or acid, containing protoxide of mercury, is treated with *hydrosulphuric acid* (HS) or *hydrosulphate of ammonia* ($NH_4S, HS$), a black precipitate of protosulphide of mercury (HgS) is thrown down, which is insoluble in dilute acids and also in excess of the hydrosulphate: it is soluble however, in aqua regia. $HgO,NO_5+HS=HgS+HO,NO_5$.

If the precipitate, after being dried, be heated alone in a tube, it is decomposed into metallic mercury and the persulphide ($HgS_2$). $2HgS=HgS_2+Hg$.

341. *Ammonia* ($NH_3$) gives a black precipitate, consisting of a basic double salt of mercury and ammonia ($NH_3,3HgO,NO_5$), which is insoluble in excess.

342. *Potash* ($KO$) produces a black precipitate of protoxide of mercury (HgO), which is insoluble in excess.

$$HgO,NO_5+KO=HgO+KO,NO_5.$$

343. (C) *Hydrochloric acid* ($HCl$), or a solution of *chloride of sodium* ($NaCl$), throws down a white precipitate of protochloride (calomel, HgCl), which is insoluble in excess.

$$HgO,NO_5+NaCl=HgCl+NaO,NO_5.$$

344. (C) Place a strip of clean *copper* in the mercurial solution, and observe the deposition of metallic mercury (324). $HgO,NO_5+Cu=Hg+CuO,NO_5$.

If the stain be rubbed, it will become bright and

silvery. Dry the stained copper, place it in a dry tube, and apply heat: the mercury sublimes, and condenses in minute globules in the upper part of the tube.

## SECTION IV.

### *Peroxide of Mercury* ($HgO_2$).

The perchloride of mercury or corrosive sublimate ($HgCl_2$), either solid or in solution, may be used.

345. (C) Heat a small fragment of the perchloride in a tube; it fuses, boils, and sublimes into the upper part of the tube. If the experiment be made on charcoal, the whole is volatilized.

346. (C) Repeat the experiment described in (336) using the perchloride instead of calomel; metallic mercury sublimes in both cases.

347. Boil a little with water, in which it readily dissolves, thus differing from the protochloride.

348. (C) Test a solution of the perchloride with a small quantity of *hydrosulphuric acid* (HS). A white precipitate is first formed, which on the addition of more of the precipitant, gradually becomes darker and ultimately black. This change of color is owing to the formation first of a double compound of sulphide and chloride of mercury ($2HgS_2,HgCl_2$), which is white; and when the hydrosulphuric acid is added in excess, the whole of the mercury is converted into the black persulphide ($HgS_2$). The precipitate is insoluble in hydrochloric and nitric acids, but is readily decomposed by aqua regia, and again converted into the perchloride.

If the persulphide be dried and cautiously sublimed in a tube, it is deposited, without decomposition, in the form of dark red crystals of cinnabar.

349. (C) *Hydrosulphate of ammonia* ($NH_4S,HS$) behaves in the same way as hydrosulphuric acid.

350. *Ammonia* ($NH_3$) throws down a white precipitate, which consists of a double compound of perchloride and amidide of mercury ($HgCl_2,Hg2NH_2$).[1]

[1] The amidides or amides, are compounds of a metal with amidogen, which is a hypothetical salt radical, supposed to consist of $NH_2$; it has,

351. (C) ***Potash*** (*KO*) gives a yellow precipitate of hydrated peroxide of mercury ($HgO_2,3HO$) which is insoluble in excess. If ammoniacal salts are present, the precipitate formed by potash is white, and consists of the same compound as that thrown down by ammonia (350).

352. (C) When *protochloride of tin* (*SnCl*) is added in small quantity the perchloride is reduced to the state of protochloride (HgCl), which separates as a white precipitate. If the salt of tin be added in excess, and the mixture boiled, the mercury is reduced to the metallic state (339).

$$HgCl_2+2SnCl=Hg+2SnCl_2.$$

353. (C) *Iodide of potassium* (*KI*) causes a most beautiful red precipitate of periodide of mercury ($HgI_2$), which surpasses even vermilion in brilliancy of color. It is readily soluble in an excess of either of the solutions.

$$HgCl_2+2KI=HgI_2+2KCl.$$

354. (C) A strip of clean *metallic copper* precipitates mercury in the metallic state (344).

355. Heat a small fragment of the red peroxide of mercury ($HgO_2$) gently in a small tube, and observe that it becomes much darker in color when hot, and reassumes its former tint on cooling. If the heat be increased to a little below redness, the oxide is decomposed into metallic mercury and oxygen, when the metal condenses in minute globules in the cool part of the tube, and the oxygen may be detected by introducing a glowing match (109).

## SECTION V.

### *Oxide of Lead* (PbO).

A solution of the acetate ($PbO,C_4H_3O_3+3Aq$) or the nitrate ($PbO,NO_5$) may be used.

356. (C) When a fragment of any of the salts of lead (except the phosphate (412) ) is heated on charcoal in the

however, never been obtained in an insulated form. See Fownes' Manual of Chemistry, p. 265.

inner flame of the blowpipe, a globule of metallic lead is formed, which is usually surrounded by a little deposit of the yellow oxide (PbO). The metallic globule will be found to be soft and malleable. In the oxidizing flame, oxide of lead forms with borax and microcosmic salt, yellowish beads, which become nearly colorless on cooling.

357. *Hydrosulphuric acid* (HS) throws down in solutions containing lead, either neutral or slightly acidified, a dense black precipitate of sulphide of lead (PbS).[1]

$$PbO,NO_5+\text{HS}=\text{PbS}+HONO_5.$$

If the sulphide be boiled with strong nitric acid, it is gradually converted into the insoluble sulphate (PbO, $SO_3$), both metal and sulphur becoming oxidized at the expense of the nitric acid.

358. *Hydrosulphate of ammonia* ($NH_4S,HS$) produces the same effect.

359. *Ammonia* ($NH_3$) and *Potash* ($KO$) throw down white precipitates, consisting of oxide of lead in combination with a small quantity of acid (basic salts).

Ammonia produces scarcely any precipitate in a solution of acetate of lead, owing to the formation of the subacetate ($3PbO,C_4H_3O_3$), which is soluble.

360. *Carbonate of potash* ($KO,CO_2$) gives a white precipitate of carbonate of lead ($PbO,CO_2$), which is insoluble in excess.

361. (C) *Sulphuric acid* ($HO,SO_3$) or a solution of *sulphate of soda* ($NaO,SO_3$), produces a white precipitate of sulphate of lead ($PbO,SO_3$), which is insoluble or nearly so in acids, but soluble in potash, and also in acetate of ammonia ($NH_4O,C_4H_3O_3$).

$$PbO,NO_5+NaO,SO_3=\text{PbO},\text{SO}_3+NaO,NO_5.$$

If the precipitate be moistened with a solution of hydrosulphate of ammonia, it is instantly blackened, owing to the formation of sulphide of lead (PbS): it is distinguished in this way from the insoluble sulphates of baryta and strontia.

362. (C) *Hydrochloric acid* ($HCl$), or a solution of *chloride of sodium* ($NaCl$), throws down a white and

[1] Under some peculiar circumstances, this reagent throws down a red precipitate in solutions of lead (see 365).

often crystalline precipitate of chloride of lead (PbCl). If the solution with the precipitate be boiled, a portion of the chloride dissolves, and is deposited again on cooling, in the form of needle-shaped crystals. If the solution of lead is dilute, the chloride does not precipitate, as it is somewhat soluble in water.

$$PbO,NO_5+HCl=\mathrm{PbCl}+HO,NO_5.$$

The chloride of lead is unaffected by an excess of ammonia.

363. (C) *Chromate of potash* ($KO,CrO_3$) gives a fine yellow precipitate of chromate of lead ($PbO,CrO_3$) which is insoluble in dilute acids, but soluble in potash. This substance is the base of the pigment known in commerce as chrome yellow.

364. (C) *Iodide of potassium* (*KI*) also gives a beautiful yellow precipitate of iodide of lead (PbI), which is rather lighter in tint than the chromate. If the iodide thus formed be boiled with water, it dissolves, and again separates on cooling, in the form of brilliant crystalline scales, which are extremely beautiful.

365. If a solution of nitrate of lead be precipitated with *hydrochloric acid*, and the filtered solution treated with *hydrosulphuric acid gas* (HS), instead of the black sulphide usually formed by that reagent in solutions of lead (357) there is produced a red precipitate, which is a chlorosulphide (3PbS,2PbCl). If the gas be passed through the solution for a length of time, however, the red compound gradually disappears, and the black sulphide (PbS) is formed.

366. (C) All the precipitates formed in the foregoing experiments, when dried, and heated on charcoal in the inner flame of the blowpipe, are decomposed, and give beads of metallic lead (356).

## SECTION VI.

### *Oxide of Copper* (CuO).

A solution of sulphate of copper ($CuO,SO_3+5Aq$) may be used.

367. (C) Heated on charcoal in the deoxidizing flame

of the blowpipe, especially if mixed with carbonate of soda, the salts of copper are reduced, and a malleable bead of the metal is obtained; the peculiar color of which may be seen on scraping off the thin coating of oxide (CuO), with which it is surrounded. In the oxidizing flame with borax or microcosmic salt, oxide of copper forms beads, which are green while hot, becoming blue on cooling.

368. *Hydrosulphuric acid* (HS) and *hydrosulphate of ammonia* ($NH_4S,HS$) throw down a black precipitate of sulphide of copper (CuS) from solutions of copper salts, whether neutral, acid, or alkaline.

$$CuO,SO_3+HS=CuS+HO,SO_3.$$

369. (C) *Ammonia* ($NH_3$), when added in small quantity, throws down a pale blue precipitate, consisting of a basis salt of copper, which immediately redissolves when the ammonia is added in excess; the solution thus formed has a beautiful deep blue color, owing to the formation of the ammonio-sulphate of copper ($2NH_3,HO,CuO,SO_3$).

370. (C) *Potash* (*KO*) produces in cold solutions of copper, a pale blue precipitate of hydrated oxide (CuO, HO). If the mixture be boiled, or if the potash be added to a hot solution, the precipitate becomes black, owing to the decomposition of the hydrated oxide at a temperature of 212°, and formation of the anhydrous black oxide (CuO). The potash must for this purpose be added slightly in excess, as otherwise the precipitate would consist of basic salt, which would not become black when boiled.

371. (C) *Ferrocyanide of Potassium* ($K_2,FeCy_3+3Aq$) gives, even in very dilute solutions, a mahogany-colored precipitate of ferrocyanide of copper ($Cu_2,FeCy_3$), which is insoluble in dilute acids.

372. (C) A piece of clean *iron*, when placed in a solution containing copper, causes a precipitation of metallic copper on its surface (324).

$$CuO,SO_3+Fe=Cu+FeO,SO_3.$$

This is an extremely delicate test, and by this means

the whole of the copper may be removed from a liquid, especially if a slight excesss of acid is present.

## SECTION VII.

### *Oxide of Silver* (AgO).

A solution of nitrate of silver ($AgO,NO_5$) may be used.

373. (C) Most of the salts of silver when exposed to light, especially when in contact with organic matter, gradually become more or less purple, and eventually nearly black, owing to partial decomposition.

374. (C) When heated on charcoal before the blowpipe, all the salts of silver are easily reduced, and a brilliant white bead of metallic silver is formed. In the oxidizing flame, oxide of silver gives with borax an opaque white bead; with microcosmic salt, the bead is yellowish by daylight, and red by candlelight.

375. *Hydrosulphuric acid* (HS) and *hydrosulphate of ammonia* ($NH_4S,HS$) throw down a black precipitate of sulphide of silver (AgS), which is insoluble in dilute acids, but soluble in boiling nitric acid.

376. (C) *Ammonia* ($NH_3$) gives a brown precipitate of oxide of silver (AgO), which is readily soluble in excess of ammonia.

Potash ($KO$) also produces the same precipitate, which is insoluble in excess.

377. (C) *Hydrochloric acid* ($HCl$) or a solution of *chloride of sodium* ($NaCl$) produces in solution of silver a white curdy precipitate of chloride of silver (AgCl), which is insoluble in water and in nitric acid, but readily soluble in ammonia, and very sparingly so in an excess either of hydrochloric acid or chloride of sodium.

$$AgO,NO_5+NaCl=\mathrm{AgCl}+NaO,NO_5.$$

If the ammoniacal solution be neutralized with nitric acid, the chloride is reprecipitated.

378. (C) *Phosphate of soda* ($2NaO,HO,PO_5+24Aq$) throws down a pale yellow precipitate of tribasic phosphate of silver ($3AgO,PO_5$), which is soluble both in nitric acid and in ammonia.

## SECTION VIII.

### *Protoxide of Tin* (SnO).

A solution of protochloride of tin (*SnCl*) may be used.

379. (C) Salts of tin, mixed with carbonate of soda ($NaO,CO_2$), and heated in the inner flame of the blowpipe, are reduced to the metallic state, and malleable globules of metallic tin are formed.

In the oxidizing flame, with borax or microcosmic salt, oxide of tin forms clear, colorless beads, unless a large quantity of the oxide is present, when the bead is sometimes opaque.

380. (C) When the neutral protosalts of tin (as the protochloride) are treated with a *large quantity of water*, they are decomposed into an acid salt (*SnCl,HCl*) which is soluble, and a basic salt (SnCl,SnO,2HO), which is insoluble: the precipitation of the latter causes the liquid to become milky.

$$3SnCl+3HO=SnCl,HCl+\mathrm{SnCl,SnO,2HO}.$$

381. (C) *Hydrosulphuric acid* (HS) gives in solutions of the protosalts of tin, either neutral or with excess of acid, a dark brown precipitate of protosulphide of tin (SnS), which is soluble in potash and in hydrosulphate of ammonia, especially if it contains an excess of sulphur (710).

382. (C) *Hydrosulphate of ammonia* ($NH_4S,HS$) also throws down the brown protosulphide, which is soluble in excess, provided a little free sulphur is present in it, which is always the case when the hydrosulphate has a yellow color (710). If the solution thus formed be neutralized with hydrochloric acid, a yellow precipitate of the persulphide ($SnS_2$) is produced, which was formed by the action of the excess of sulphur in the hydrosulphate upon the protosulphide. $SnS+S=SnS_2$.

383. *Ammonia* ($NH_3$) gives a bulky white precipitate of hydrated oxide of tin (SnO,HO), which is insoluble in excess.

$$SnCl+NH_3+2HO=\mathrm{SnO,HO}+NH_4Cl.$$

384. *Potash* (*KO*) also produces a white precipitate

of hydrated oxide ($SnO,HO$), which redissolves in an excess of the alkaline solution.

If a concentrated solution of the oxide in potash be boiled, the protoxide is converted into a mixture of peroxide ($SnO_2$) and metallic tin; the first remains in solution, and the latter precipitates.

$$2SnO = SnO_2 + Sn.$$

385. *Carbonate of potash* ($KO,CO_2$) also throws down the hydrated oxide, which is insoluble in excess.

386. (C) *Terchloride of gold* ($AuCl_3$) causes in solutions of the protosalts of tin, a dark purple precipitate, which has long been known as *purple of Cassius;* its composition appears to be ($2(SnO,SnO_2)+AuO_2SnO_2+6HO$). For the success of this experiment, it is necessary that both solutions be exceedingly dilute.

## SECTION IX.

### *Peroxide of Tin* ($SnO_2$).

A solution of the perchloride ($SnCl_2$) may be used.

387. (C) Salts of the peroxide of tin behave in the same manner before the blowpipe as those of the protoxide (379).

388. (C) *Hydrosulphuric acid* ($HS$) gives a yellow precipitate of persulphide of tin ($SnS_2$), which is soluble in solution of potash.

389. (C) *Hydrosulphate of ammonia* ($NH_4S,HS$) also throws down the yellow persulphide, which is readily soluble in excess.

390. *Ammonia* ($NH_3$) and *potash* ($KO$) throw down a bulky white precipitate of hydrated peroxide of tin. ($SnO_2,HO$), which is soluble in an excess of the precipitant, especially when potash is used, forming a compound called stannate of potash ($KO,SnO_2$), in which the peroxide of tin appears to play the part of an acid.

The hydrated peroxide, when thus formed by precipitation with potash, is readily soluble both in potash and nitric acid, in which respect it differs from that formed by the action of nitric acid on metallic tin, though both yield the same results when analyzed (392).

391. (C) Pour a few drops of *nitric acid* upon a small fragment of metallic tin in a test-tube, and observe the intense action which immediately takes place. The nitric acid ($NO_5$) is decomposed by the affinity of the tin for a portion of its oxygen; the white hydrated peroxide of tin ($SnO_2,HO$) is formed; and nitric oxide ($NO_2$) and some of the other oxides of nitrogen are given off. A little ammonia ($NH_3$), also, is at the same time formed, owing to the decomposition of water by the tin; the hydrogen combining with some of the nitrogen derived from the nitric acid.

$$7Sn+3NO_5+3HO=7SnO_2+2NO_2+NH_3.$$

392. Heat the hydrated oxide formed in the last experiment, first with nitric acid and afterwards in a solution of potash, and observe that it is quite insoluble in both, thus differing from that formed by potash (390).

Fig. 74.

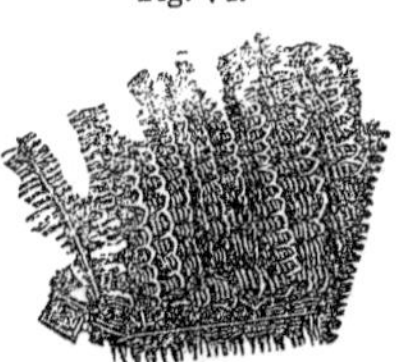

Crystals of Metallic Tin.

393. (C) If a piece of clean *zinc* be placed in a solution of perchloride of tin, the tin is separated in the metallic state, in the form of a beautiful feathery crystals; some of which are so minute, as to look like an amorphous spongy mass, but when examined with the microscope, appear as multitudes of brilliant and beautifully formed crystalline tufts (Fig. 74).

$$SnCl_2+2Zn=Sn+2ZnCl.$$

## SECTION X.

### *Oxide of Bismuth* ($Bi_2O_3$).

A solution of the chloride ($Bi_2Cl_3$) may be used.

394. (C) Mix a concentrated solution of the chloride with a considerable quantity of *water*, which causes a white precipitate of oxychloride of bismuth ($Bi_2Cl_3, 2Bi_2O_3$).

A similar decomposition takes place when solutions of many of the soluble salts of bismuth are diluted with much water. The precipitates thus formed are usually distinguishable from those produced under the same circumstances in solutions of antimony, by being insoluble in tartaric acid (332). They dissolve easily, however, in acetic acid.

395. (C) When the salts of bismuth are mixed with carbonate of soda, and heated in the reducing flame of the blowpipe, small globules of the metal are formed, which break with crystalline fracture when struck with a hammer. In the oxidizing flame, with borax or microcosmic salt, oxide of bismuth forms a yellowish bead, which becomes nearly colorless on cooling.

396. *Hydrosulphuric acid* (HS), and *hydrosulphate of ammonia* ($NH_4S,HS$) throw down from solutions of bismuth, which do not contain a large excess of free acid, a black precipitate of sulphide of bismuth ($Bi_2S_3$), which is insoluble in dilute acid and potash, but soluble in hot nitric acid.

397. *Ammonia* ($NH_3$) and *potash* ($KO$) give a white precipitate of hydrated oxide ($Bi_2O_3,3HO$), which is insoluble in an excess of the precipitant.

398. *Carbonate of potash* ($KO,CO_2$) gives a bulky white precipitate of subcarbonate of bismuth ($Bi_2O_3,CO_2$), which is insoluble in excess.

399. (C) When oxide of bismuth is heated, it turns yellow, and becomes colorless again on cooling.

---

# CHAPTER VI.

## ACTION OF REAGENTS WITH THE INORGANIC ACIDS.

400. THE inorganic acids, which are enumerated in paragraph (179), may be conveniently divided into three classes, according to their behavior with chloride of barium and nitrate of silver, thus:—

*Class I.*—Acids which are precipitated by a solution of chloride of barium.

| | |
|---|---|
| Sulphuric ($HO,SO_3$). | Carbonic ($CO_2$). |
| Phosphoric ($PO_5$). | Silicic ($SiO_3$). |
| Boracic ($BO_3$). | |

*Class II.*—Acids which are unaffected by chloride of barium, but which are precipitated by a solution of nitrate of silver.

Hydrochloric ($HCl$), Hydriodic ($HI$), and Hydrosulphuric ($HS$).

*Class III.*—Those which are not precipitated either by chloride of barium or nitrate of silver.

Nitric ($NO_5$) and Chloric ($ClO_5$).

## SECTION I.

### *Sulphuric acid* ($HO,SO_3$).

401. Mix a few drops of strong sulphuric acid or oil of vitriol with about an equal quantity of water in a test-tube, and observe the heat evolved.

402. (C) Place a small bit of wood or paper in a test-tube, and pour upon it a few drops of oil of vitriol: the organic matter is decomposed, and black carbonaceous matter is formed.

403. (C) Add a few drops of a solution of *chloride of barium* ($BaCl$), or *nitrate of baryta* ($BaO,NO_5$) to one of sulphate of soda ($NaO,SO_3$): a heavy white precipitate of sulphate of baryta ($BaO,SO_3$) is thrown down, which is insoluble in hydrochloric acid.

$$BaCl+NaO,SO_3=BaO,SO_3+NaCl.$$

404. (C) *Acetate of lead* ($PbO,C_4H_3O_3+3Aq$) throws down in solutions containing sulphuric acid, a dense white precipitate of sulphate of lead ($PbO,SO_3$) which is insoluble in dilute acids, but sparingly soluble in strong sulphuric and hydrochloric acids (691). It is soluble also in potash and acetate of ammonia ($NH_4O, C_4H_3O_3$).

$$PbO,C_4H_3O_3+NaO,SO_3=PbO,SO_3+NaO,C_4H_3O_3.$$

405. (C) Mix a little dry sulphate of soda ($NaO,SO_3$) or some other sulphate, with black flux, and heat it on

platinum wire in the reducing flame of the blowpipe: the oxygen both of the soda and acid is removed, and sulphide of sodium (NaS) remains.

$$NaO,SO_3+2C=NaS+2CO_2.$$

406. (C) Place the bead formed in the last experiment in a test-tube, and moisten it with a little dilute sulphuric acid: hydrosulphuric acid (HS) is given off, which may be recognized by its odor: or by putting into the tube a strip of paper moistened with a solution of acetate of lead, which will be blackened, owing to the formation of sulphide of lead (PbS) (438).

$$NaS+HO,SO_3=NaO,SO_3+HS.$$

## SECTION II.

### *Phosphoric Acid (tribasic)* ($3HO,PO_5$).

A solution of common tribasic phosphate of soda ($2NaO,HO,PO_5+24Aq$) may be used.[1]

407. *Chloride of barium* ($BaCl$) throws down a white precipitate of phosphate of baryta ($2BaO,HO,PO_5$), which is soluble in hydrochloric acid.

$$2BaCl+2NaO,HO,PO_5=2NaCl+2BaO,HO,PO_5.$$

408. *Chloride of calcium* ($CaCl$) gives a white precipitate of phosphate of lime ($8CaO,3PO_5$), which readily dissolves in a slight excess of hydrochloric acid.[2]

409. (C) *Sulphate of magnesia* ($MgO,SO_3+7Aq$) causes a white precipitate of phosphate of magnesia ($2MgO,HO,PO_5$), if the solution is tolerably strong. If a little ammonia or carbonate of ammonia, however, be present in the solution, the double phosphate of ammonia and magnesia ($2MgO,NH_4O,PO_5+12Aq$), is formed, which being much more insoluble than the phosphate of magnesia, is precipitated in more dilute solutions, and is consequently a more delicate test. It separates as a

[1] The monobasic ($HO,PO_5$) and bibasic phosphoric acid ($2HO,PO_5$), being rarely met with in analysis, are omitted.

[2] See note to (217).

granular crystalline precipitate, and is readily soluble in excess of acid. If the double phosphate be heated to redness, the ammonia and water volatilize, and the anhydrous phosphate of magnesia ($2MgO,PO_5$) is left behind (206).

410. (C) *Nitrate of Silver* ($AgO,NO_5$) throws down a pale yellow precipitate of tribasic phosphate of silver ($3AgO,PO_5$), which is soluble both in ammonia and nitric acid (310).

$$3(AgO,NO_5)+2NaO,HO,PO_5=3AgO,PO_5+2(NaO,NO_5)HO,NO_5.$$

411. (C) Heat a small fragment of common tribasic phosphate of soda before the blowpipe; when cool, dissolve it in water, and add to the solution a few drops of *nitrate of silver*. Instead of the yellow tribasic phosphate of silver ($3AgO,PO_5$) being thrown down as before, a white granular precipitate of the bibasic phosphate ($2AgO,PO_5$) is produced. This is owing to the tribasic phosphate of soda having been converted into the bibasic phosphate ($2NaO,PO_5$) by the expulsion of the equivalent of basic water, when heated.

412. (C) *Acetate of lead* ($PbO,C_4H_3O_3+3Aq$) gives a white precipitate of phosphate of lead ($3PbO,PO_5$), which is soluble in nitric acid. If this precipitate be collected on a filter, dried, and heated before the blowpipe, it fuses into a semi-transparent bead, which on cooling becomes very distinctly crystalline. This test is decidedly characteristic, not only on account of the crystalline structure of the bead, but from the circumstance that the phosphate, unlike the other salts of lead, is not easily reduced to the metallic state when heated in the inner flame.

413. If *perchloride of iron* ($Fe_2Cl_3$) be added to a solution of a phosphate acidified with a little hydrochloric acid, and subsequently mixed with solution of *acetate of potash* ($KO,C_4H_3O_3$), the phosphoric acid is thrown down in combination with peroxide of iron ($2Fe_2O_3,3HO,3PO_5$). If this phosphate of iron be digested with hydrosulphate of ammonia, it is decomposed; sulphide of iron (FeS) is formed, and the phosphoric acid remains in solution in combination with ammonia. The phosphate of iron may be still more completely decomposed

by first dissolving it in a slight excess of hydrochloric acid, and nearly neutralizing with ammonia before the addition of the hydrosulphate.

## SECTION III.

### *Boracic Acid* ($BO_3$).

A solution of borax ($NaO,2BO_3+10Aq$) may be used.

414. (C) A solution of borax turns turmeric paper brown, thus resembling an alkali or alkaline carbonate.

Boracic acid in solution produces the same effect, though in a less degree.

415. *Chloride of barium* ($BaCl$) throws down a white precipitate of borate of baryta ($BaO,2BO_3$), which is readily soluble in hydrochloric acid.

$$BaCl+NaO,2BO_3=BaO,2BO_3+NaCl.$$

416. *Nitrate of silver* ($AgO,NO_5$) gives a white precipitate of borate of silver ($AgO,BO_3$), which is soluble both in ammonia and nitric acid.

417. (C) If *strong sulphuric acid* ($HO,SO_3$) be added to a concentrated solution of a borate, the boracic acid which is displaced, separates in combination with water in the form of crystalline scales.

$$NaO,2BO_3+HO,SO_3=NaO,SO_3+2BO_3+HO.$$

418. (C) If borax or any other borate be moistened with a little *sulphuric acid*, and the mixture treated with *alcohol*, the boracic acid is dissolved, and communicates a green color to the flame when it is burnt. This is probably owing to a little of the boron (B) being deoxidized by the burning spirit, and recombining with oxygen as it comes in contact with the air at the edge of the flame.

## SECTION IV.

### *Carbonic Acid* ($CO_2$).

The physical and some of the chemical properties of carbonic acid have been already noticed (18, &c.).

419. (C) The carbonates when treated with a *free acid*

as hydrochloric, are decomposed, and the carbonic acid, being gaseous when uncombined, escapes with effervescence.

$$CaO,CO_2+HCl=CaCl+HO+CO_2.$$

It may be distinguished from other gaseous acids by being inodorous. When a substance such as marble is tested in this way for carbonic acid, it is generally advisable to drench it with water; if this is not done, small bubbles of common air, which at first adhere to the solid substance, gradually escape, and may lead an inexperienced person to suppose that effervescence is taking place.

420. (C) When carbonic acid is passed into *lime water* (*CaO*), it causes a white precipitate of carbonate of lime ($CaO,CO_2$), most of which redissolves if the gas is passed through for a length of time, owing to the formation of the bicarbonate of lime ($CaO,2CO_2$), which is soluble in water.

Fig. 75.

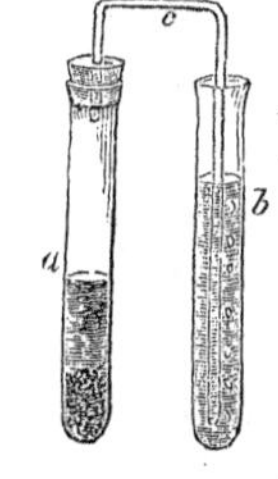


This experiment is best made in a test-tube *a* (Fig. 75), to which is connected, by means of a perforated cork, a bent tube, *c*. A small lump of marble is put into *a*, and the tube *b* half filled with lime-water: dilute hydrochloric acid is then poured upon the marble, and the bent tube attached, which conducts the liberated carbonic acid into the lime-water, which it immediately renders turbid.

421. *Chloride of barium* ($BaCl$) and *chloride of calcium* ($CaCl$) throw down a white precipitate of carbonate of baryta ($BaO,CO_2$) or of lime ($CaO,CO_2$), which readily dissolves with effervescence in dilute hydrochloric acid.

422. (C) *Subacetate of lead* ($3PbO,C_4H_3O_3$) is an extremely delicate test for carbonic acid, with which it forms a white precipitate of carbonate of lead ($PbO,CO_2$).

423. Most of the carbonates, except those of the alkalies, are decomposed when strongly heated; in which case the oxide or the reduced metal is left (122).

## SECTION V.

### *Silicic Acid* ($SiO_3$).

There are two modifications of silicic acid, one of which is soluble, and the other insoluble.

424. (C) Add a little strong *hydrochloric acid* ($HCl$) to a concentrated solution of silicate of potash ($KO,SiO_3$), and warm the mixture: a bulky precipitate separates, which is soluble in potash, while a portion remains dissolved in the acid solution; this is the soluble modification (probably a definite hydrate) of silicic acid.

$$KO,SiO_3+HCl=KCl+HO+SiO_3.$$

425. (C) Evaporate to dryness the solution with the precipitate, formed in the last experiment, and observe, that on again treating the residue with hydrochloric acid, the silicic acid remains undissolved; in this state it is almost insoluble also in cold alkaline solutions. Thus we find, that when the soluble modification of silicic acid is evaporated to dryness, it is converted into the insoluble modification.

426. (C) Mix a little dry silicic acid, or an insoluble silicate, in the fine powder with dry carbonate of soda ($NaO,CO_2$), and fuse it, for about ten minutes, on a platinum wire before the blowpipe; treat the bead with dilute hydrochloric acid, and observe that the insoluble silicic acid has been changed, by the fusion with the alkali, into the soluble modification. If the solution thus obtained be evaporated to dryness, the silicic acid again becomes insoluble.

427. (C) When pure silicic acid is fused with carbonate of soda before the blowpipe, a transparent colorless bead of silicate of soda is formed, while carbonic acid is expelled.

$$NaO,CO_2+SiO_3=NaO,SiO_3+CO_2.$$

In this experiment a small quantity only of the soda should be used, as it forms an opaque bead when added in excess.

## SECTION VI.

### *Hydrochloric Acid* ($HCl$).

(*Chlorine, in combination with hydrogen or a metal.*)

A solution of chloride of sodium ($NaCl$), or the dilute acid, may be used.

428. *Chloride of barium* gives no precipitate. If, however, it be added to strong hydrochloric acid, the chloride of barium will be precipitated unchanged, as it is insoluble in the strong acid.[1]

429. (C) *Nitrate of silver* ($AgO,NO_5$) throws down a white curdy precipitate of chloride of silver (AgCl), which is insoluble in nitric acid, but readily soluble in ammonia.

$$AgO,NO_5+NaCl=\mathrm{AgCl}+NaO,NO_5.$$

If the ammoniacal solution be neutralized with nitric acid, the chloride is again precipitated.

This precipitate, like most of the salts of silver, becomes purple on exposure to light.

430. (C) *Acetate of lead* ($PbO,C_4H_3O_3+3Aq$) gives a white precipitate of chloride of lead (PbCl) in tolerably strong solutions of chlorides; if the precipitate be boiled with a little water, it dissolves, and separates again on cooling, in the form of needle-shaped crystals.

$$PbO,C_4H_3O_3+NaCl=\mathrm{PbCl}+NaO,C_4H_3O_3.$$

431. (C) When mixed with *nitric acid*, and warmed, hydrochloric acid dissolves *gold leaf*, forming terchloride of gold ($AuCl_3$).

$$3HCl+NO_5+\mathrm{Au}=AuCl_3+\mathrm{NO_2}+3HO.$$

[1] This circumstance must be remembered when testing hydrochloric acid with chloride of barium, with a view to ascertaining whether it contains traces of sulphuric acid; in which case it is necessary to dilute the acid before testing.

## SECTION VII.

### *Hydriodic Acid (KI).*

(*Iodine in combination with hydrogen or a metal.*)

A solution of iodide of potassium (*KI*) may be used.

432. *Chloride of barium* gives no precipitate with hydriodic acid.

433. (C) *Nitrate of silver* ($AgO,NO_5$) gives a pale straw-colored precipitate of iodide of silver (AgI), which gradually becomes purple when exposed to the light. It is nearly insoluble in nitric acid, and considerably less soluble in ammonia than the chloride (429).

434. (C) *Perchloride of mercury* ($HgCl_2$) gives a brilliant red precipitate of periodide of mercury ($HgI_2$), which dissolves in an excess either of the perchloride or of the iodide of potassium.

$$HgCl_2+2KI=HgI_2+2KCl.$$

435. (C) *Starch* ($C_{12}H_{10}O_{10}$) forms with iodine, even in highly dilute solutions, a dark purple precipitate of iodide of starch. If the iodine is in a state of combination, as in iodide of potassium or hydriodic acid, it is necessary to liberate it before applying the starch; which is readily done by adding a drop or two of nitro-hydrochloric acid (698), or a solution of chlorine; if nitric acid is employed, a portion of its oxygen combines with the hydrogen or metal with which the iodine was in combination, forming water, or a metallic oxide.

$$3KI+NO_5=3KO+NO_2+I.$$

If chlorine be used, it forms with the hydrogen or metal, hydrochloric acid or a metallic chloride; iodine being liberated in either case.

$$KI+Cl=KCl+I.$$

The starch may be applied either in solution or as a paste; or, what is often more convenient, strips of paper or cotton may be impregnated with the solution, dried, and kept for use (750).

436. (C) If iodide of potassium, or any other metallic iodide, in the solid state, be heated with a little strong *sulphuric acid* ($HO,SO_3$), both compounds are decomposed; sulphurous acid ($SO_2$) and potash ($KO$) are formed, and the iodine is set free.

$$KI+2(HO,SO_3)=KO,SO_3+SO_2+2HO+I.$$

A portion of the latter sublimes in the form of a beautiful violet-colored vapor, which condenses in the upper part of the tube, and is highly characteristic.

If the quantity of iodine liberated is so small that the color of the vapor is not perceptible, it may readily be detected by suspending a bit of paper or cotton moistened with a solution of starch, which will instantly be turned purple (435).

437. *Dilute sulphuric acid* when added to the solution of an iodide, also causes its decomposition, especially if the mixture be boiled, setting free a little iodine, which gives a pale yellowish color to the solution, and causes a purple precipitate with solution of starch.

## SECTION VIII.

### *Hydrosulphuric Acid* (HS).[1]

(*Sulphur, in combination with hydrogen or a metal.*)

438. (C) Most of the metallic sulphides are decomposed when treated with hydrochloric acid, in which case hydrosulphuric acid (HS) is given off, and may be recognized by its disagreeable odor, resembling that of rotten eggs.

Add a little dilute *hydrochloric* or *sulphuric acid* to a small fragment of sulphide of iron (FeS) in a test-tube; hydrosulphuric acid is immediately evolved; and if a strip of paper, moistened with a solution of acetate of lead, be held over the open end, it will be blackened, owing to the formation of the black sulphide of lead (PbS). The gas may also be passed into a solution of the acetate, in the manner shown in (420), when it will throw down the black sulphide (357).

[1] Called also *Sulphuretted Hydrogen.*

$$FeS + HCl = FeCl + HS.$$

439. (C) When heated with *nitric acid* ($NO_5$), the metallic sulphides are decomposed: the metal is oxidized at the expense of a portion of the nitric acid; orange fumes of nitrous acid and nitric oxide being given off; while the sulphur separates as a whitish powder, which gradually collects into yellowish lumps, and is eventually dissolved, owing to its conversion into sulphuric acid, also at the expense of the nitric acid.

$$CuS + 2NO_5 = CuO,SO_3 + NO_2 + NO_4.$$

440. (C) The soluble sulphides, or hydrosulphates, are also decomposed by acids, with evolution of hydrosulphuric acid. Add a little *dilute hydrochloric acid* to a drop or two of hydrosulphate of ammonia ($NH_4S, HS$); hydrosulphuric acid is given off, while muriate of ammonia ($NH_4Cl$) remains in solution, and a little free sulphur is at the same time deposited, which had before been dissolved in the hydrosulphate, causing a white precipitate.

$$NH_4S,HS + HCl = NH_4Cl + 2HS.$$

When hydrosulphate of ammonia is first prepared, it is colorless, but a portion of the hydrosulphuric acid is gradually decomposed by the affinity of the atmospheric oxygen for its hydrogen, with which it combines to form water, while at the same time the equivalent of sulphur is set free ($HS + O = HO + S$); the latter dissolves in the hydrosulphate, giving it a yellow color. It is this sulphur which is precipitated on the addition of an excess of acid to the hydrosulphate (710).

441. *Chloride of barium* gives no precipitate with hydrosulphuric acid, or the hydrosulphates.

442. *Nitrate of silver* ($AgO,NO_5$) gives a black precipitate of sulphide of silver ($AgS$), which is soluble in hot nitric acid.

443. (C) *Acetate of lead* ($PbO,C_4H_3O_3 + 3Aq$) throws down in solutions of hydrosulphuric acid or the hydrosulphates, a black precipitate of sulphide of lead ($PbS$), which is converted into sulphate of lead ($PbO,SO_3$) by boiling with nitric acid, which furnishes oxygen to both elements.

$$PbO,C_4H_3O_3+HS=PbS+HO,C_4H_3O_3.$$
$$PbS+4O=PbO,SO_3.$$

444. (C) Before the blowpipe, the sulphides are readily decomposed; the sulphur is driven off and burns with a blue flame, forming sulphurous acid ($SO_2$), which may be recognized by its odor, which is well known as that of burning sulphur.

## SECTION IX.

### *Nitric Acid* ($HO,NO_5$).

Nitrate of potash ($KO,NO_5$), both solid and in solution, may be used.

445. *Chloride of barium* ($BaCl$) gives no precipitate in solutions of the nitrates. If it be added, however, to strong nitric acid, a white precipitate will be produced, consisting merely of the undecomposed chloride; which, though soluble in water, is insoluble in the strong acid.[1]

446. *Nitrate of silver* causes no precipitate in solutions of the nitrates.

447. (C) If a small fragment of nitrate of potash be placed on ignited charcoal, vivid deflagration takes place, owing to the rapid combination of the carbon with oxygen, which it abstracts from the nitre; carbonic acid ($CO_2$) is thus formed, which combines with the potash previously in combination with the nitric acid.

$$KO,NO_5+C=KO,CO_2+NO_3.$$

448. (C) When a nitrate is heated with a little strong *sulphuric acid* ($HO,SO_3$), it is decomposed; and if copper filings are added to the mixture, the copper becomes oxidized at the expense of the liberated nitric acid; nitric oxide and nitrous acid are given off, forming orange-colored fumes, which are very characteristic.

$$KO,NO_5+3Cu+4(HO,SO_3)=3(CuO,SO_3)+KO,SO_3+4HO+NO_2.$$

449. (C) Add a few drops of strong *sulphuric acid* to a solution of a nitrate in a test-tube, and when the mix-

[1] This must be borne in mind when testing the purity of nitric acid.

ture is cold (401), drop in a small crystal of protosulphate of iron ($FeO,SO_3+7Aq$). When nitric acid is present, a brown compound is formed round the crystal, consisting of protoxide of iron ($FeO$) in combination with nitric oxide ($NO_2$), while the other three equivalents of oxygen combine with another portion of the protoxide, forming sesquioxide of iron ($Fe_2O_3$), which is dissolved by the sulphuric acid as sesquisulphate ($Fe_2O_3,3SO_3$).

$$10(FeO,SO_3)+4(HO,SO_3)+KO,NO_5=3(Fe_2O_3,3SO_3)+KO,SO_3 \\ +4(FeO,SO_3),NO_2+HO.$$

If the mixture is heated, the brown compound is decomposed, and the color disappears.

450. (C) If a little *hydrochloric acid* be added to a solution containing nitric acid or a nitrate, the mixture has the property of dissolving gold leaf, owing probably to the liberation of free chlorine, which acts on the metal. The terchloride of gold ($AuCl_3$) thus formed, gives the solution a yellowish color.

$$KO,NO_5+2HCl=KCl+2HO+NO_4+Cl.$$

451. Strong nitric acid has the property of turning many nitrogenous organic compounds yellow; a fact of of which most chemists have unintentionally convinced themselves while experimenting with nitric acid, by the troublesome yellow stains it leaves on the fingers; the cuticle being converted into a compound called xanthoproteic acid. ($2HO,C_{34}N_4H_{24}O_{12}$.)

452. (C) If a nitrate be mixed with a little *sulphuric acid*, and warmed with a drop or two of *sulphate of indigo*, the blue color of the latter disappears, owing to the conversion of the indigo into colorless oxidized compounds.

## SECTION X.

### *Chloric Acid* ($ClO_5$).

Chlorate of potash ($KO,ClO_5$), both solid and in solution, may be used.

453. *Neither chloride of barium* nor *nitrate of silver* produce any precipitate in solutions of the chlorates.

454. (C) Heat a small fragment of the chlorate in a test-tube with the flame of a spirit lamp; it is decomposed, and if the heat is continued long enough, the whole of the oxygen is given off. Chloride of potassium (KCl) remains behind.

$$KO,ClO_5 = KCl + 6O.$$

The presence of oxygen may be proved by introducing an ignited match into the tube.

455. (C) Dissolve the residue of chloride of potassium formed in the last experiment, in water, and test the solution with *nitrate of silver* (429). The formation of a chloride after the application of heat is the best proof that the acid is chloric and not nitric.

456. (C) When placed on ignited charcoal, or when heated with organic substances, the chlorates deflagrate even more violently than the nitrates. On this account very small fragments only of the chlorate should be used.

457. (C) Place a small fragment of chlorate of potash ($KO,ClO_5$) in a test-tube, and pour upon it a few drops of strong sulphuric acid ($HO,SO_3$) *taking especial care not to warm the mixture*, as it is liable to explode with violence when heated. The chlorate is decomposed, sulphate of potash ($KO,SO_3$) and perchlorate of potash ($KO,ClO_7$) are formed, together with peroxide of chlorine ($ClO_4$), which gives the mixture a yellowish color, and escapes in the form of a greenish gas.

$$3(KO,ClO_5) + 2(HO,SO_3 = 2(KO,SO_3) + KO,ClO_7 + 2ClO_4 + 2HO.$$

458. Repeat experiments 448, 449, and 452, using chlorate of potash instead of the nitrate, and compare the results with those obtained with the latter.

---

## CHAPTER VII.

### ORGANIC ACIDS.

459. The organic acids which are enumerated in paragraph 179, may be divided into three classes, ac-

cording to their behavior with *chloride of calcium* and *perchloride of iron;* thus:—

*Class I.*—Organic acids which are, under certain circumstances, precipitated by a solution of chloride of calcium.

| | |
|---|---|
| Oxalic ($HO,C_2O_3$). | Citric ($3HO,C_{12}H_6O_{12}$). |
| Tartaric ($2HO,C_8H_4O_{10}$). | Malic ($2HO,C_8H_4O_8$). |

*Class II.*—Those which are unaffected by chloride of calcium, but are thrown down by a solution of perchloride of iron.

| | |
|---|---|
| Succinic ($HO,C_4H_2O_3$). | Benzoic ($HO,C_{14}H_5O_3$). |

*Class III.*—Those which are not precipitated either by chloride of calcium or by perchloride of iron.

| | |
|---|---|
| Acetic ($HO,C_4H_3O_3$). | Formic ($HO,C_2HO_3$). |

## SECTION I.

### *Oxalic Acid* ($HO,C_2O_3$).

460. Oxalic acid is readily soluble in water and in alcohol.

461. When crystallized oxalic acid ($HO,C_2O_3+3Aq$) is heated in a tube, a portion volatilizes unchanged, while a part is decomposed.

462. (C) All the salts of the organic acids are decomposed at a red heat; and when the base is an alkali, or alkaline earth, a carbonate of the base is formed. This decomposition is almost always attended with a deposition of charcoal, and consequent blackening; but in the case of the oxalates scarcely any blackening takes place, the oxalic acid being almost wholly resolved into carbonic acid ($CO_2$) and carbonic oxide ($CO$); the latter of which escapes, while the carbonic acid combines with the base.

$$KO,C_2O_3=KO,CO_2+CO.$$

463. (C) Heat a little oxalate of lime ($CaO,C_2O_3$) to low redness for a few moments on platinum foil, and observe that the decomposition takes place almost without blackening. Place the fragment in a test-tube, and moisten it with dilute hydrochloric acid, when the effer-

vescence will show the presence of carbonic acid (419, 122).

464. (C) *Chloride of calcium* ($CaCl$), when added to solutions containing oxalic acid, either free or in combination with a base, causes, even in highly dilute solutions, a copious white precipitate of oxalate of lime ($CaO,C_2O_3+2Aq$), which is readily soluble in hydrochloric acid, and slightly so in an excess of oxalic acid, so that the addition of ammonia favors the precipitation in an acid solution.

$$CaCl+NH_4O,C_2O_3=CaO,C_2O_3NH_4Cl.$$

The presence of ammoniacal salts does not interfere with the formation of this precipitate.

465. (C) A solution of *sulphate* ($CaO,SO_3$) or any other salt *of lime*, causes the same precipitate ($CaO,C_2O_3+2Aq$), even in very dilute solutions. Lime-water also does the same.

466. *Perchloride of iron* ($Fe_2Cl_3$) gives no precipitate in solutions of oxalic acid or the oxalates, unless they are tolerably concentrated.

467. *Nitrate of silver* ($AgO,NO_5$) throws down a white precipitate of oxalate of silver ($AgO,C_2O_3$), which is soluble both in nitric acid and ammonia. If the precipitate be dried, and heated on platinum foil, it is dispersed with a slight puff, leaving a residue of metallic silver.

468. (C) When oxalic acid or an alkaline oxalate is warmed with strong *sulphuric acid* ($HO,SO_3$), it is decomposed into carbonic acid ($CO_2$) and carbonic oxide ($CO$), while the basic water or the alkali combines with the sulphuric acid (33).

$$KO,C_2O_3+HO,SO_3=KO,SO_3+HO+CO_2+CO.$$

The two gases escape with effervescence, and if a taper be applied as they issue from the tube, the carbonic oxide burns with a pale blue flame, combining with an additional equivalent of oxygen from the air, and becoming carbonic acid (41).

## SECTION II.

### *Tartaric Acid* ($2HO,C_8H_4O_{10}$).

469. Tartaric acid is soluble both in water and in alcohol.

470. (C) Heat a small crystal of the acid in a tube; it at first fuses, and is afterwards decomposed, with deposition of carbon, and consequent blackening. A peculiar and characteristic odor is at the same time emitted.

471. Fold a small fragment of bitartrate of potash ($KO,HO,C_8H_4O_{10}$) in platinum foil, and heat it to redness before the blowpipe or over a spirit lamp. The tartaric acid is thus decomposed, and the carbonate of potash ($KO,CO_2$) is at the same time formed. Place the fragment in a test-tube, and add a few drops of dilute hydrochloric acid, when it will effervesce, showing the presence of carbonic acid (419).

472. (C) *Chloride of calcium* ($CaCl$) throws down in neutral solutions containing tartaric acid, a white precipitate of tartrate of lime ($2CaO,C_8H_4O_{10}$), which is soluble in a cold solution of potash; if the potash solutions be heated, however, the tartrate of lime separates as a bulky precipitate, but redissolves as the solution cools.

473. *Lime-water* ($CaO$) causes in neutral solutions a white precipitate of tartrate of lime ($2CaO,C_8H_4O_{10}$), which is soluble in an excess of acid. The presence of ammoniacal salts prevents the formation of this precipitate, though if the mixture be allowed to stand a few hours, the tartrate of lime gradually crystallizes out.

474. *Sulphate of lime* ($CaO,SO_3$) gives no precipitate at first, even in neutral solutions of tartrates; but if allowed to stand, tartrate of lime gradually crystallizes.

475. *Salts of potash* cause the formation of bitartrate of potash ($KO,HO,C_8H_4O_{10}$), which separates from the solution in the form of a granular precipitate, soluble in an excess of alkali, and most of the inorganic acids. If the tartaric acid is present as a neutral tartrate, the bisulphate of potash ($KO,HO,2SO_3$) should be employed for testing it. In dilute solutions the separation of the precipitate is hastened by agitating the liquid with a glass rod, when lines of minute crystals will be deposited on the sides of the glass wherever the rod has rubbed against it (184, 186).

476. *Perchloride of iron* ($Fe_2Cl_3$) gives no precipitate with tartaric acid or the tartrates.

477. (C) *Acetate of lead* ($PbO,C_4H_3O_3+3Aq$) throws down a white precipitate of tartrate of lead ($2PbO,C_8H_4O_{10}$), which when washed clean, is readily soluble in ammonia.

478. Tartaric acid and the tartrates, when present in solutions of the persalts of iron, prevent the precipitation of the hydrated peroxide ($Fe_2O_3,3HO$) when ammonia or potash are added (280). This is owing to the formation of double tartrates of iron and the alkali, which are soluble in water, and are not decomposed by an excess of the latter. Tartaric acid also prevents the precipitation of alumina, protoxide of manganese, and some other oxides, under similar circumstances.

## SECTION III.

### *Citric Acid* ($3HO,C_{12}H_6O_{12}$).

479. Citric acid is soluble in water and in alcohol.

480. (C) When heated in a tube, citric acid at first melts, and is subsequently decomposed, emitting pungent fumes, which may be distinguished by their smell from those formed by tartaric acid under similar circumstances. A carbonaceous residue remains in the tube.

481. (C) *Chloride of calcium* ($CaCl$), when added to solutions of neutral citrates, gives a white precipitate of citrate of lime ($3CaO,C_{12}H_6O_{12}$), which is insoluble in potash, but soluble in muriate of ammonia. If the ammoniacal solution be boiled, the citrate of lime reprecipitates. Free citric acid gives no precipitate with this test.

482. (C) *Lime-water* ($CaO$) fails to produce a precipitate in a cold solution, but if the mixture be boiled, citrate of lime is thrown down, being less soluble in hot water than in cold.

483. *Perchloride of iron* ($Fe_2Cl_3$) gives no precipitate.

484. *Acetate of lead* ($PbO,C_4H_3O_3+3Aq$) throws down a white precipitate of citrate of lead ($3PbO,C_{12}H_6O_{12}$), which when washed, is only very slightly soluble in ammonia, thus differing from tartaric acid (477).

485. Citric acid and the soluble citrates, when present

in solutions containing peroxide of iron, alumina, and some other metallic oxides, prevent their precipitation by ammonia, owing to the formation of soluble double salts.

486. Citric acid when heated with strong *sulphuric acid*, is decomposed; carbonic acid and carbonic oxide are given off with effervescence, and after some time, sulphurous acid ($SO_2$) is formed, and the mixture becomes dark colored.

## SECTION IV.

### *Malic Acid* ($2HO,C_8H_4O_8$).

487. Malic acid dissolves freely both in water and alcohol.

488. (C) When malic acid is cautiously heated in a tube, it is decomposed into two new acids; maleic acid ($2HO,C_8H_2O_6$), which being volatile, sublimes and condenses in the upper part of the tube; and fumaric acid ($HO,C_4HO_6$) which remains at the bottom. If the heat is allowed to rise higher than 400° or 500°, further decomposition takes place, and the mass is carbonized.

489. *Chloride of calcium* ($CaCl$) gives no precipitate, since the malate of lime ($2CaO,C_8H_4O_8$) is soluble in water; the addition of alcohol, however, immediately causes it to precipitate.

490. *Lime water* ($CaO$) gives no precipitate with malic acid or the malates.

491. *Perchloride of iron* ($Fe_2Cl_3$) causes no precipitate, as the malate of iron is soluble in water.

492. (C) *Acetate of lead* ($PbO,C_4H_3O_3+3Aq$) throws down a white precipitate of malate of lead ($PbO,HO,C_8H_4O_8$). If acetate of lead in solution be allowed to stand for a day or two on the precipitate, it is gradually converted into beautiful tufts of silky crystals. If the precipitate be well washed, and, while suspended in water, heated over a lamp, it will be found to melt into a resin-like mass at the temperature of boiling water.

493. Like tartaric and citric acids, malic acid and the soluble malates prevent the precipitation of peroxide of iron and some other metallic oxides by the alkalies (478, 485).

494. When heated with *oil of vitriol* ($HO,SO_3$), malic acid is decomposed and carbonized, sulphurous acid ($SO_2$) being at the same time given off.

SECTION V.

*Succinic Acid* ($HO,C_4H_2O_3$).

495. Succinic acid is soluble both in water and alcohol.

496. (C) When the pure acid is heated in a tube, it volatilizes entirely, leaving no carbonaceous residue, and crystallizes in the upper part of the tube. The common acid met with in commerce is seldom pure, and usually leaves a slight residue.

497. *Chloride of calcium* ($CaCl$) gives no precipitate with succinic acid or the succinates.

498. (C) *Perchloride of iron* ($Fe_2Cl_3$) throws down a bulky light brown precipitate of persuccinate of iron ($Fe_2O_3,2C_4H_2O_3$) from perfectly neutral solutions containing succinic acid. This precipitate is soluble in acids, and is decomposed by ammonia, which removes the greater part of the acid.

499. *Acetate of lead* ($PbO,C_4H_3O_3+3Aq$) gives a white precipitate of succinate of lead ($PbO,C_4H_2O_3$), which is soluble in acid solutions, and is decomposed into a basic salt by ammonia.

500. (C) When treated with a mixture of *chloride of barium, ammonia, and alcohol*, solutions containing succinic acid give a white precipitate of succinate of baryta ($BaO,C_4H_2O_3$).

SECTION VI.

*Benzoic Acid* ($HO,C_{14}H_5O_3$).

501. Benzoic acid is scarcely soluble in cold water, but rather more so in hot; it is readily soluble in alcohol.

502. (C) When heated in a tube, it sublimes and condenses in the form of beautiful needle-shaped crystals: the vapor has a peculiar aromatic odor, and causes an unpleasant sensation in the throat, inducing coughing.

503. *Chloride of calcium* ($CaCl$) gives no precipitate in solutions of benzoic acid, the benzoate of lime being soluble in water.

504. *Perchloride of iron* ($Fe_2Cl_3$) gives in neutral solutions, a light yellowish brown precipitate of perbenzoate of iron ($Fe_2O_3,3C_{14}H_5O_3$), which is soluble in acids, and like the succinate, is decomposed by ammonia (498).

505. *Acetate of lead* ($PbO,C_4H_3O_3+3Aq$) throws down a white precipitate of benzoate of lead ($PbO,C_{14}H_5O_3$) in solutions of benzoate of potash or of soda, but not in a solution of the free acid, or of benzoate of ammonia.

506. (C) A mixture of *chloride of barium, ammonia, and alcohol*, gives no precipitate with benzoic acid and the benzoates, thus differing from succinic acid (500).

507. (C) When the solution of an alkaline benzoate, as benzoate of ammonia ($NH_4O,C_{14}H_5O_3$), is treated with strong *sulphuric* or *hydrochloric acid*, it is decomposed, and the liberated benzoic acid, being almost insoluble in water, is precipitated in the form of a white crystalline precipitate; while the sulphate or muriate of ammonia remains in solution.

$$NH_4O,C_{14}H_5O_3+HO,SO_3=NH_4O,SO_3+HO,C_{14}H_5O_3.$$

## SECTION VII.

### *Acetic Acid* ($HO,C_4H_3O_3$).

508. Acetic acid is soluble in all proportions in water: it dissolves also in alcohol.

509. (C) When heated, it volatilizes readily, leaving, if pure, no residue; the fumes have an exceedingly pungent odor, resembling that of vinegar, which owes its active properties to the acetic acid which it contains.

510. *Chloride of calcium*, and the other salts of lime, give no precipitate with acetic acid or the acetates.

*Perchloride of iron* also gives no precipitate, but changes the color of the solution to a deep reddish brown.

511. (C) *Nitrate of silver* ($AgO,NO_5$) causes in neutral solutions, a white precipitate of acetate of silver ($AgO, C_4H_3O_3$), which if the mixture is heated, partially dis-

solves, and recrystallizes on cooling; it is soluble in ammonia.

511 a. (C) Protonitrate of mercury ($HgO,NO_5$) gives on agitation a precipitate of acetate of mercury ($HgO, C_4H_3O_3$), which separates in the form of white silky crystalline scales.

$$HgO,NO_5+KO,C_4H_3O_3=HgO,C_4H_3O_3+KO,NO_5.$$

512. (C) When an acetate is mixed with dilute *sulphuric acid*, and gently warmed, it is decomposed; a sulphate of the base being formed, while the acetic acid is set free, and may be recognized by its odor.

$$KO,C_4H_3O_3+HO,SO_3=KO,SO_3+HO,C_4H_3O_3.$$

513. (C) If a mixture of an acetate with *dilute sulphuric acid* be distilled (536), and the distilled liquid boiled with an excess of *oxide of lead* ($PbO$), the liberated acetic acid combines with a portion of the oxide, forming subacetate of lead ($3PbO,C_4H_3O_3$); which, having an alkaline reaction with test paper, may be recognized by its turning the yellow color of turmeric paper brown.

514. (C) When acetic acid or an acetate is warmed with strong *sulphuric acid* ($HO,SO_3$), acetic ether ($C_4H_5O,C_4H_3O_3$) is formed, which volatilizes, and may be known by its peculiar and refreshing odor.

## SECTION VIII.

### *Formic Acid* ($HO,C_2HO_3$).

515. Formic acid is readily soluble both in water and alcohol.

516. (C) When heated, it volatilizes entirely, giving off fumes of a penetrating disagreeable odor.

517. *Chloride of calcium* gives no precipitate with formic acid or the formiates.

518. *Perchloride of iron* also gives no precipitate.

519. *Nitrate of silver* gives in strong and neutral solutions of the formiates, a white precipitate of formiate of silver ($AgO,C_2HO_3$) which shortly becomes of a darker color, owing to decomposition and the liberation of

metallic silver. If the mixture be boiled, this reduction takes place immediately.

A similar decomposition takes place with protonitrate of mercury.

520. (C) When warmed with *dilute sulphuric acid*, the formiates are decomposed: the formic acid volatilizes, and may be known by its odor.

521. (C) If formic acid or a formiate be heated with *strong sulphuric acid*, it is resolved into water and carbonic oxide (CO). $C_2HO_3=2CO+HO$.

The carbonic oxide gas escapes with effervescence, and burns with a pale blue flame if a light be applied to the mouth of the tube.

# PART III.

## QUALITATIVE ANALYSIS OF SUBSTANCES, THE COMPOSITION OF WHICH IS UNKNOWN.

---

## CHAPTER I.

### PRELIMINARY EXAMINATION, ETC.

522. When a substance is presented for examination, with a view to ascertaining its chemical composition, it is obvious that it would be extremely tedious if we were to begin by applying indiscriminately the tests for the various metals and acids, until we happened to meet with one which gave a characteristic reaction; and although such a method might occasionally succeed in the examination of substances consisting merely of one base and acid, it would certainly be found wholly inefficient under less favorable circumstances, as when two or more bases and acids are mixed together, which would mask or neutralize each other's behavior with the different reagents employed. Hence the necessity of a well-devised plan of proceeding, in which, by means of a few simple experiments, we are enabled to obtain some considerable insight into the nature of the substance under examination.

523. Before beginning what may be called the *real analysis* of a substance, it is generally advisable to make a few preliminary experiments upon it, in order to ascertain the class of compounds to which it belongs, and whether the usual mode of analysis will be likely to succeed with it: this may be called the *preliminary examination*.

We will first suppose that a solid substance has been given to the student for examination. If it is a liquid, he may pass on to paragraph (533).

SECTION I.

*Preliminary Examination of Solids.*[1]

524. Observe whether the substance is CRYSTALLINE or AMORPHOUS: whether it is HOMOGENEOUS THROUGHOUT, or composed of different ingredients, such as can be distinguished either with the naked eye, or with the assistance of a lens.

Note any peculiarity of FORM or COLOR; and observe whether or not it possesses METALLIC LUSTRE.

525. TAKE ITS SPECIFIC GRAVITY (145). A knowledge of this is frequently of great service, especially in the case of minerals, when by a reference to a table of specific gravities,[2] we are able at once to guess what it is, and to strike out of the list of possibilities a large number of substances which it might more or less resemble in external appearance.

526. PLACE A FRAGMENT OF THE SUBSTANCE IN A SMALL TUBE OF HARD GERMAN GLASS, CLOSED AT ONE END: HEAT IT FIRST OVER A LAMP, AND AFTERWARDS IN THE FLAME OF THE BLOWPIPE. Observe (*a*) WHETHER IT APPEARS TO UNDERGO ANY CHANGE: if it does not, we infer that the substance contains no water or organic matter; that it is not readily fusible; and that no volatile substances are present.

(*b*) DOES IT FUSE? and if so, does it continue fluid as long as the heat is applied, or does it, after a short time, solidify while still in a heated state? If it solidifies while hot, it had probably undergone what is called the watery fusion, or melted in its water of crystallization, which is gradually expelled, condensing in the upper part of the tube. If the substance fuses without any

[1] I cannot too strongly insist once more on the importance of making careful and accurate notes of all the experiments and observations which are made; they are not only often absolutely necessary for reference in the subsequent stages of the analysis, but the practice is also of the greatest value to the student, in cultivating habits of correct observation and facility of expression; besides at the same time impressing the facts more strongly on his recollection (6).

[2] Such a table may be found in Dr. Thomson's *Mineralogy*, vol. i p. 710.

other apparent change, it probably contains either an alkali or an alkaline earth.

(*c*) DOES IT WHOLLY VOLATILIZE? If it does, of course no fixed matter is present, and we are thus enabled to lessen the circle of our inquiry very considerably.

(*d*) Perhaps A PORTION VOLATILIZES, LEAVING A FIXED RESIDUE: hence we infer that the substance under examination is probably a mixture of two or more substances.

(*e*) When the substance is volatile (either wholly or in part), observe whether THE VAPOR CONDENSES IN THE COOL PART OF THE TUBE; and if so, whether the matter deposited is solid (crystalline or amorphous) or liquid; if the latter, is it neutral or otherwise to test paper? Is the vapor COMBUSTIBLE? Has it any CHARACTERISTIC SMELL, as of ammonia, or of burning sulphur, or of arsenic (301)?

(*f*) Does the substance under examination BLACKEN WHEN HEATED? If it does, we may infer that some organic matter is present: and if, by continuing the heat with excess of air on a piece of platinum foil, the blackness disappears, we may pronounce with certainty that such is the case. If the burnt mass, when cold, effervesces on being moistened with DILUTE HYDROCHLORIC ACID ($HCl$), while the substance in its original state does not, we may infer that an organic acid was present, which, when heated, is converted into carbonic acid, forming a carbonate (462, 471). In this case, too, it is highly probable that the base with which the organic acid was in combination, is either an alkali (potash or soda), or an alkaline earth (lime, magnesia, baryta, or strontia); as otherwise the newly-formed carbonate would probably have been decomposed, leaving either a metallic oxide or reduced metal (423).

(*g*) In case of carbonization, observe whether any CHARACTERISTIC SMELL is given off during the decomposition (470, 480), and also whether the vapor which is formed is neutral, acid, or alkaline to test paper: if alkaline, it is probable either that ammonia was present, or that nitrogen was contained in the organic matter.[1]

[1] When an organic substance containing nitrogen is heated, ammonia ($NH_3$) is almost invariably formed during the destructive decomposition.

It is not to be considered certain that, because a substance does not char when heated, no organic matter is present, since many organic substances volatilize without decomposition.

527. HEAT A FRAGMENT OF THE SUBSTANCE ON CHARCOAL IN THE INNER FLAME OF THE BLOWPIPE. In this experiment, some of the appearances already obtained by heating in a tube (526), will probably be repeated, such as charring, volatilization, &c. Observe,

(*a*) Whether THE SUBSTANCE FUSES, either easily or only after prolonged application of the flame. If fusion takes place speedily, and especially if the fused mass is absorbed by the charcoal, it is probable that potash or soda is present: if the former, the flame may be tinged with a violet color (187).

(*b*) If the substance FUSES AND BOILS, and after a short time SOLIDIFIES while still under the influence of the heat, the fusion was probably owing to the presence of water of crystallization (526 *b*).

(*c*) If the substance is INFUSIBLE (either without or subsequent to the watery fusion), and remains on the charcoal in the form of a colorless infusible mass, the substance is probably an alkaline earth, silica, oxide of zinc, or alumina. If an alkaline earth, it will probably radiate an intense white light while ignited. The white infusible mass may then be moistened with a solution of nitrate of cobalt, and again heated; when if it becomes blue, alumina may be suspected (245); if green, oxide of zinc (261); and if pale pink, magnesia (209). If it is silica, it will fuse into a clear colorless bead with carbonate of soda, effervescing at the same time (427).

(*d*) In case a BEAD OF REDUCED METAL, or a COLORED INFUSIBLE RESIDUE, is formed on the charcoal, it should be mixed with carbonate of soda ($NaO,CO_2$), and again heated as before on charcoal in the deoxidizing flame. If tin, copper, silver, or gold are present, a bead of the metal will be formed without any incrustation on the charcoal.[1] If iron, cobalt, or nickel, are present, they

[1] The white or brownish ash which is always formed when charcoal is burnt (which consists of the incombustible matter of the charcoal), must not be mistaken for an incrustation derived from the substance under examination.

will be reduced to the metallic state, but instead of fusing into a bead, will be mixed up with the carbonate of soda (being infusible except at a higher temperature), giving the bead a gray opaque appearance.

(*e*) If a WHITE DEPOSIT IS FORMED ON THE CHARCOAL ROUND THE BEAD OF METAL (or indeed without any metallic bead), it is probably owing to the presence of zinc or antimony.[1] If zinc, the oxide while hot is yellowish, becoming white on cooling (260).

(*f*) If a YELLOW OR A BROWN DEPOSIT IS FORMED, either lead, bismuth, or cadmium, may be supposed to be present.

528. WARM A FRAGMENT OF THE SUBSTANCE IN A TUBE WITH STRONG SULPHURIC ACID ($HO,SO_3$):—

(*a*) If EFFERVESCENCE OCCURS,[2] it is probably owing to the escape of some volatile acid, which is displaced by the sulphuric acid. If the gas thus liberated has no smell, carbonic acid may be suspected (419). If it smells of hydrosulphuric acid, the substance was probably a sulphide (438), in which case sulphurous acid ($SO_2$) would probably at the same time be formed. If the smell resembles that of nitric or nitrous acid, the presence of a nitrate may be inferred, especially if orange-colored fumes are evolved on the addition of clean copper filings (448). If the disengaged gas is greenish yellow, with a smell somewhat resembling chlorine, it is probably owing to the presence of a chlorate (457).

(*b*) If organic matter is present (which will have been already ascertained (526*f*)), the escape of gas may be due to the carbonic acid or other gas formed by the action of the sulphuric acid upon it; and consequently the presence of a volatile acid is not proved by this experiment, when organic matter is present.

(*c*) If the substance undergoes NO APPARENT CHANGE by the action of sulphuric acid, the absence of all these compounds may be inferred.

529. The next point to be ascertained in the preliminary examination, is as to the solubility of the sub-

[1] See note in preceding page.

[2] Care must be taken not to mistake the bubbles of common air, which often escape from the surface of a solid substance when it is treated with a liquid, for true effervescence.

stance in water and other solvents. Place five or ten grains of the pounded substance in a test-tube, and treat it with a little distilled water (at first cold, and afterwards boiled if the substance does not dissolve), and observe whether it is wholly or partially soluble, or whether it is absolutely insoluble. This is known by evaporating a drop of the clear liquid (filtered if necessary) on platinum foil, when, if anything is dissolved, it will be left as a residue; which, if abundant, indicates that the substance is copiously soluble; and if slight, that it is only sparingly so.[1]

(*a*) If NOTHING IS DISSOLVED by the water, we thus prove the absence of all soluble compounds.

(*b*) If it WHOLLY DISSOLVES, we prove the absence of all insoluble compounds.

(*c*) If it PARTIALLY DISSOLVES, it is either a sparingly soluble substance, or a mixture of soluble and insoluble matters: the addition of more water will show which of these is the case. If it is a mixture, the insoluble portion may be separated by filtration from the solution, for further examination (530).

530. IF THE SUBSTANCE, OR ANY PORTION OF IT, IS FOUND TO BE INSOLUBLE, OR BUT VERY SPARINGLY SOLUBLE, IN WATER, IT MUST BE TREATED WITH DILUTE HYDROCHLORIC ACID ($HCl$), (EXCEPT IT BE A METAL, *see* 531,) AND IF NECESSARY BOILED. Whether or not anything dissolves, may be ascertained by evaporating a drop of the clear liquid on platinum foil (529). If EFFERVESCENCE OCCURS, it is probably owing to the escape of a gaseous acid, which may sometimes be identified by the smell or color (528 *a*). If the smell of chlorine be given off, it may be owing to the presence of a peroxide, as of manganese ($MnO_2$).

$$MnO_2+2HCl=MnCl+2HO+Cl.$$

In case the whole, or any portion, of the substance prove insoluble in hydrochloric acid, it must be sepa-

[1] It is always necessary to test by experiment, whether the distilled water used in these experiments, is itself perfectly pure, and free from dissolved matter: if such were not the case, a residue would of course be left, even though the substance under examination were insoluble (753).

rated by filtration, and retained for further examination (532).

531. IF THE SUBSTANCE IS A METAL (KNOWN BY ITS METALLIC LUSTRE, &c.), IT MUST BE TREATED WITH STRONG NITRIC ACID, AND IF NECESSARY BOILED.

(*a*) If NO APPARENT ACTION TAKES PLACE, the metal is probably gold or platinum: and if it be found that nothing has dissolved, the absence of all the common, easily oxidizable metals may be inferred.

(*b*) If the METAL IS ACTED UPON, AND A WHITE PRECIPITATE IS AT THE SAME TIME FORMED, which is found to be insoluble in water, it is probable that antimony or tin is present (391); and if, besides the formation of the white precipitate, some of the metal is dissolved (known by evaporating a drop of the clear liquid on platinum foil), the presence of some other metal, soluble in nitric acid, may be relied on.

(*c*) If the metal DISSOLVES ENTIRELY, the absence of gold, antimony, and tin, may be inferred.

532. THE MATTER (IF ANY) WHICH PROVED INSOLUBLE IN HYDROCHLORIC ACID (530), IS NOW TREATED, FIRST WITH STRONG NITRIC ACID; AND THEN, IF IT RESISTS SOLUTION, WITH NITROHYDROCHLORIC ACID (698), AND IF NECESSARY BOILED. If insoluble in this, it is probably one of the insoluble silicates, sulphates, or chlorides, and will have to be afterwards examined (578, 623).

## SECTION II.

### *Preliminary Examination of Liquids.*

533. When the substance given for examination is liquid, a drop or two should be evaporated on platinum foil, to ascertain whether or not it contains any fixed matter in solution. If such is the case, a small quantity of the liquid is to be evaporated to dryness in a basin, and the residue examined according to the directions given above for solid substances (524, 526 et seq.). Towards the end of the evaporation, when the residue is nearly dry, and a pellicle of solid matter is formed on the surface, it is very liable to "spurt," and project small portions of the substance out of the basin (644); this is

best avoided by moderating the heat, and by constantly stirring with a glass rod, so as to prevent the formation of the pellicle.

While the evaporation is going on, the following experiments may be commenced with the solution.

534. TAKE ITS SPECIFIC GRAVITY (148).

535. TEST THE SOLUTION WITH LITMUS AND TURMERIC PAPER, to ascertain whether it is neutral or otherwise.

(*a*) If NEUTRAL, the absence of free acids and alkalies, and of acid salts, may of course be considered certain. It is probable, also, that the only salts present are those of the alkalies or alkaline earths, as the solutions of most other salts have a feebly acid reaction.

(*b*) If it has an ACID REACTION (known by its reddening blue litmus paper), it is owing to the presence of either an uncombined acid, an acid salt, or a soluble salt of one of the heavy metals, many of which have a feebly acid reaction. To ascertain which of these is present, pour a little of the solution into a test-tube, and stir it with a glass rod, the end of which is moistened with a solution of carbonate of potash ($KO,CO_2+2Aq$); if this cause a precipitate, the acid reaction is probably owing to the presence of a metallic salt: while, if the solution remains clear, a free acid or an acid salt is probably the cause.

(*c*) If the solution has an ALKALINE REACTION (known by its turning turmeric paper brown), it is probably owing to the presence of a free alkali or alkaline earth, one of the alkaline carbonates, or an alkaline sulphide. In this case we are enabled to exclude at once all oxides which are insoluble in alkaline solutions: and if alkaline carbonates are present, none of the alkaline earths can exist in solution, since they would be thrown down as insoluble carbonates.

536. If the solvent liquid is supposed from its TASTE OR SMELL, to be other than water, it may be necessary to insulate it from the solid matter it contains for examination. This is best done, if the liquid is volatile, by distillation in a small retort (61), or if the quantity of liquid is minute, the distillation may be effected in two small tubes, as shown in Fig. 76, *a* being the retort,

and *b* the receiver; the latter may, if necessary, be kept cool by immersion in cold water. The liquid should be poured down a long tube-funnel (Fig. 77) to avoid

Fig. 76. Fig. 77.

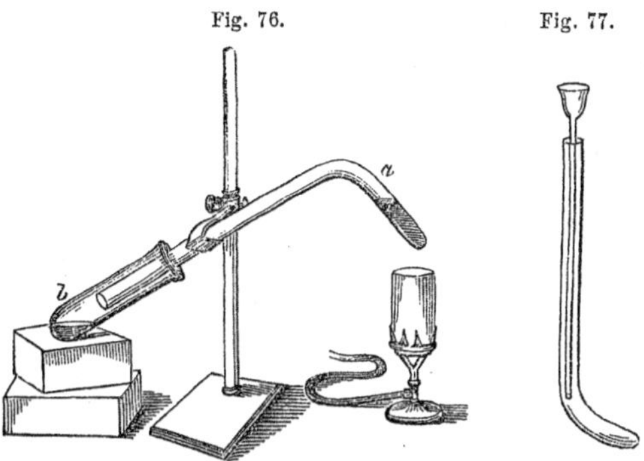


Tube Distillation.

soiling the long limb of *a*. The distilled liquid may then be examined as to its taste, smell, specific gravity, boiling-point, &c.

537. When the substance to be examined is liquid, containing solid matter in suspension, the latter is to be separated by filtration, and the solid and liquid portions examined separately, according to the directions given in paragraph 524 et seq. and 533 et seq.

## SECTION III.

### *Actual Analysis.*

### Introductory Remarks.

538. Having learnt from the preliminary experiments just described the general nature of the substance under examination, together with its degree of solubility, &c., we proceed to the actual analysis by means of liquid tests, with a view to ascertaining the exact constituents of which it is composed.

We will first, for the sake of simplicity, and leaving entirely out of sight all the rarer substances (179), describe the processes to be followed in the analysis of simple salts which are known to contain only one metallic oxide or base, combined with one acid; as, for example, sulphate of potash ($KO,SO_3$); or a binary compound of a metal with a nonmetallic body (or *haloid* salt), such as chloride of calcium ($CaCl$); and first those which are readily soluble in water. It is usual to determine the base first, and when that is done, the student may pass on to (556), and commence testing for the acid.

539. When the presence of any metal or acid is indicated by the action of a reagent employed in qualitative analysis, it is always necessary to confirm our supposition by applying other tests; as it is rarely the case that a single test is sufficiently decided in its results, to render the presence of a metal absolutely certain. The student, therefore, when he is led to infer from the result of an experiment, that a certain substance is present, should refer to the action of other reagents on the particular metal or acid in question; when he will have no difficulty, by applying two or three of the most characteristic tests to some of the original solution, in proving his supposition to be correct or otherwise.

---

## CHAPTER II.[1]

### QUALITATIVE ANALYSIS OF A SIMPLE SALT, CONTAINING ONE BASE AND ONE ACID, WHICH IS READILY SOLUBLE IN WATER (529).

#### SECTION I.

*Examination of the Base of the Salt.*

540. Having made a tolerably strong solution of the salts, a little of the solution is treated with a drop or two of DILUTE HYDROCHLORIC ACID (*HCl*). If this causes a WHITE PRECIPITATE, it is probably owing to the presence

[1] The student will find in the Appendix a list of salts, &c., which may be taken for practice in qualitative analysis. He may first examine a few of each kind with the assistance of the book, until he finds himself tolerably familiar with the processes; after which he may try them without reference to the printed directions.

of either *Lead*, *Silver*, or *Protoxide of Mercury*, the chlorides of which, being more or less insoluble, are precipitated as soon as formed. In order to distinguish between them, a portion of the liquid with the precipitate, is supersaturated with AMMONIA ($NH_3$).

(*a*) If this DISSOLVES THE PRECIPITATE, the base is probably *Oxide of Silver* (377).

(*b*) If the PRECIPITATE BECOMES DARK COLORED, the base is probably *Protoxide of Mercury* (338).

(*c*) If the PRECIPITATE REMAINS UNALTERED by the ammonia, the base is probably *Oxide of Lead* (362).

In either case, a portion of the original solution should be tried with some of the most characteristic tests for the suspected metal (539).

*Hydrosulphuric Acid Test.*

541. If no precipitate is caused by the hydrochloric acid, the acidified portion of the solution is mixed with one of HYDROSULPHURIC ACID ($HS$), or the gas may be passed through it (701) until it smells perceptibly. If THIS CAUSES NO CHANGE, the student may pass on to (547); but if A PRECIPITATE IS PRODUCED, the base is thus shown to be one of these in the fourth class (179), since none of the others are precipitated from an acidified solution by hydrosulphuric acid.

542. If THE PRECIPITATE IS BLACK, the base is either oxide of *Lead*, oxide of *Copper*, oxide of *Bismuth*,[1] or peroxide of *Mercury*. To prove which of these it is, add to separate portions of the original solution in a test-tube, the following tests, until one of them is found to indicate the metal present.

(*a*) Add a little DILUTE SULPHURIC ACID ($HO,SO_3$). If this causes a WHITE PRECIPITATE, the base is probably *Oxide of Lead*, the white precipitate being in that case sulphate of lead ($PbO,SO_3$) (361). To confirm this, add some of the other tests for lead (356, 363, 364).

(*b*) If THE SULPHURIC ACID GIVES NO PRECIPITATE, add to another portion a solution of AMMONIA ($NH_3$); if this causes at first a LIGHT BLUE PRECIPITATE, which on the

[1] Bismuth is seldom met with in insoluble compounds since most of its salts are insoluble, or only very sparingly soluble, in water.

addition of ammonia in excess, redissolves, forming a deep rich BLUE SOLUTION, the base is *Oxide of Copper* (369). (Confirm 371, 372.)

(*c*) To another portion of the original solution, add a few drops of a solution of POTASH (*KO*); if this causes a YELLOW PRECIPITATE, the base is *Peroxide of Mercury* (351). (Confirm 346, 353, 354.)

(*d*) If none of these tests succeed, a little of the solution should be EVAPORATED NEARLY TO DRYNESS WITH HYDROCHLORIC ACID, and then added to a considerable quantity of water in a test-tube; if a WHITE PRECIPITATE is produced, the base is probably *Oxide of Bismuth* (394). (Confirm 395, 397.)

543. If THE PRECIPITATE CAUSED BY HYDROSULPHURIC ACID IS BROWN, the base is probably *Protoxide of Tin* (381). (Confirm 379, 382, 386.)

544. If THE PRECIPITATE CAUSED BY HYDROSULPHURIC ACID IS YELLOW, either the base is *Peroxide of Tin* (388), or one of the oxides of *arsenic* is present (307).[1] To determine which of these it is, add to a portion of the original solution, a few drops of dilute AMMONIA; if this cause a WHITE PRECIPITATE, the base is probably *Peroxide of Tin* (390) (confirm 387, 393); while if the SOLUTION REMAINS CLEAR, the yellow sulphide is probably that of *Arsenic*.[2] (Confirm 303, 312.)

545. If THE COLOR OF THE PRECIPITATE THROWN DOWN BY HYDROSULPHURIC ACID IS ORANGE, the base is probably *Oxide of Antimony* (329). (Confirm 332, 334.)

546. If THE PRECIPITATE WITH HYDROSULPHURIC ACID IS WHITE, the base is probably *Peroxide of Iron*, sulphur being in that case precipitated (278). (Confirm 280, 282.)

[1] Although both of the oxides of arsenic ($AsO_3$ and $AsO_5$) have acid properties, they are best included here among the bases, on account of their behavior with hydrosulphuric acid.

[2] When in the examination of a salt, the precipitate with hydrosulphuric acid is found to be owing to the presence of arsenic, we must still seek for the base by further experiments, since both the oxides of arsenic are acids. When the substance is soluble in water, the base in combination with the arsenious or arsenic acid will probably be found to be one of the alkalies, since the arsenites and arseniates of all the other metallic oxides are insoluble in water.

*Hydrosulphate of Ammonia Test.*

547. When hydrosulphuric acid causes no precipitate, it may be inferred that no metal of the fourth class (179) is present, and that the metal contained in the salt belongs consequently to one of the three other classes. Add a little MURIATE OF AMMONIA ($NH_4Cl$)[1] to a portion of the original solution, and then a few drops of dilute AMMONIA, unless the solution was quite neutral; in which case the addition of ammonia is unnecessary, its use being to prevent the presence of any excess of acid, which might interfere with the action of the hydrosulphate of ammonia (440).

548. ADD NOW TO THE NEUTRAL OR SLIGHTLY AMMONIACAL SOLUTION, HYDROSULPHATE OF AMMONIA. If this causes NO PRECIPITATE, none of the metals of the third class can be present, and the student may pass on to (553). If A PRECIPITATE APPEARS, however, the base is thus shown to be one of those included in the third class: viz. *Alumina* oxide of *Chromium*, oxide of *Zinc*, protoxide of *Manganese*, *Protoxide of Iron*, *Peroxide of Iron*, oxide of *Nickel*, or oxide of *Cobalt*.

549. If THE PRECIPITATE IS BLACK, the base is either protoxide or peroxide of *Iron*, oxide of *Nickel*, or oxide of *Cobalt*. To distinguish between them, add to a fresh portion of the solution a little caustic POTASH.

(*a*) If this causes a DULL PALE GREEN PRECIPITATE, which on exposure to the air becomes RUST COLORED,[2] the base is *Protoxide of Iron* (273). (Confirm 276.)

(*b*) If it throws down a RUST COLORED PRECIPITATE, the base is probably *Peroxide of Iron* (281). (Confirm 282.)

(*c*) If the precipitate caused by potash is PALE GREEN, which does not become brown by exposure to the air, the base is probably *Oxide of Nickel* (288). (Confirm 287, 291, 292.)

[1] Muriate of ammonia is here added to prevent the precipitation of any magnesia that may be present (200); which, as it does not belong to the third class of metals, might cause confusion.

[2] When it is expected that a change of color will be caused by exposing a precipitate to the air, the best way is to pour a little of the precipitate with the solution containing it, on a piece of filtering-paper; when it will come more completely in contact with the air than when allowed to remain in the test-tube.

(*d*) If the precipitate is LIGHT BLUE, changing to dirty pink when boiled, the base is probably *Oxide of Cobalt* (295). (Confirm 296, 299.)

550. If THE PRECIPITATE CAUSED BY HYDROSULPHATE OF AMMONIA IS FLESH-COLORED, becoming brown by exposure to the air, the base is probably *Protoxide of Manganese* (263). (Confirm 264, 267.)

551. If THE PRECIPITATE THROWN DOWN BY THE HYDROSULPHATE IS WHITE, the base is either *Alumina* or *Oxide of Zinc*. To distinguish between them, add to a fresh portion of the original solution a little DILUTE AMMONIA.

(*a*) If this causes a WHITE PRECIPITATE, which is READILY SOLUBLE in excess of ammonia, the base is *Oxide of Zinc* (255). (Confirm 260, 261.)

(*b*) If, on the contrary, the white precipitate thrown down by ammonia is INSOLUBLE IN EXCESS, the base is *Alumina* (241). (Confirm 245.)

552. If THE PRECIPITATE CAUSED BY THE HYDROSULPHATE IS GREEN, the base is probably *Oxide of Chromium* (247). (Confirm 248, 251, 252.)

*Carbonate of Soda Test.*

553. In case neither hydrosulphuric acid nor hydrosulphate of ammonia produce any precipitate, we know that no metal of the third or fourth class can be present, and that the base we are in search of must consequently belong either to the first or second class.

ADD A SLIGHT EXCESS OF CARBONATE OF SODA ($NaO, CO_2$) TO A PORTION OF THE ORIGINAL SOLUTION: if this causes NO PRECIPITATE, the base does not belong to Class II, and the student may pass on to (555). If, on the contrary, a WHITE PRECIPITATE IS PRODUCED, the base is one of those included in the second class, viz. *Magnesia*, *Lime*, *Baryta*, or *Strontia*.

554. To determine which of these it is, add to a little of the original solution in a neutral and concentrated state, a few drops of a solution of SULPHATE OF SODA as long as it causes any precipitate.

(*a*) If this causes NO PRECIPITATE, even after standing

[1] The sulphate of lime requires for its solution 500 times its weight of water; so that even a concentrated solution of this salt contains the lime in a highly diluted state.

a few minutes, the base is probably *Magnesia* (205) (confirm 206, 209); or *Lime* in the state of sulphate of lime ($CaO,SO_3$) (215).[1] (Confirm 216, 217, 219.)

(*b*) If, on the contrary, A PRECIPITATE IS PRODUCED, the base is either *Lime*, *Baryta*, or *Strontia*: to distinguish between them, filter the mixture, and test the clear filtered liquid with OXALATE OF AMMONIA. If this causes a WHITE PRECIPITATE, the base is probably *Lime* (216). (Confirm 215, 218.)

(*c*) If THE OXALATE CAUSES NO IMMEDIATE PRECIPITATE IN THE FILTERED SOLUTION, add to a little of the original solution, a solution of SULPHATE OF LIME. If this causes AN IMMEDIATE PRECIPITATE, the base is probably *Baryta* (225). (Confirm, 227, 228.)

(*d*) If THE PRECIPITATE DOES NOT APPEAR AT FIRST, on the addition of the sulphate of lime, BUT GRADUALLY SEPARATES after some little time, the base is probably *Strontia* (233). (Confirm 236.)

555. If neither hydrosulphuric acid, hydrosulphate of ammonia, nor carbonate of soda, produce any precipitate, the base is one of the first class viz., *Potash*, *Soda*, or *Ammonia*. To ascertain which of these it is,

(*a*) Add to a portion of the dry salt, or of the concentrated solution, in a test-tube, a little caustic POTASH ($KO$), and boil: if the SMELL OF AMMONIA is perceptible, and if the vapor produces dense white fumes when a rod, moistened with hydrochloric acid, is held near the mouth of the tube, the base is *Ammonia* (195). (Confirm 192, 194.)

(*b*) If it is not ammonia, add a little BICHLORIDE OF PLATINUM to the concentrated solution; if this causes a YELLOW CRYSTALLINE PRECIPITATE, either immediately or after standing a short time, the base is *Potash* (185). (Confirm 186, 187.)

(*c*) If NO PRECIPITATE APPEARS, and if the solution from the last experiment, on evaporating spontaneously, DEPOSITS YELLOW NEEDLE-SHAPED CRYSTALS, which are

[1] After determining the base of a salt which we know to be soluble in water, it is of course unnecessary, in the subsequent examination, to look for any acid that forms with the base an insoluble salt. (See Table of Solubilities, in the Appendix.)

readily soluble in water, the base is *Soda* (188). (Confirm 189, 190.)

SECTION II.

*Examination for the Acid.*[1]

556. Having ascertained the base of the salt under examination, we next proceed to discover the acid with which it is combined; and here we will, as before, for the sake of simplicity, leave out of sight all the rarer ones, and confine ourselves to those inorganic acids which are most commonly met with in analysis, viz.,

| | |
|---|---|
| Sulphuric ($HO,SO_3$). | Hydrochloric ($HCl$). |
| Phosphoric ($PO_5$). | Hydriodic ($HI$). |
| Boracic ($BO_3$). | Hydrosulphuric ($HS$). |
| Carbonic ($CO_2$). | Nitric ($HO,NO_5$). |
| Silicic ($SiO_3$). | Chloric ($HO,ClO_5$). |

557. A portion of the original solution, which for this purpose should be tolerably concentrated, is first treated with DILUTE SULPHURIC ACID. If NO APPARENT CHANGE takes place, or if MERELY A PRECIPITATE IS PRODUCED, the student may pass on to (558); but if EFFERVESCENCE ENSUES, the acid is probably either *Carbonic*, or *Hydrosulphuric*.

(*a*) If the gas evolved is INODOROUS, the acid is probably *Carbonic* (419). (Confirm 420, 421.)

(*b*) If the gas has a SMELL RESEMBLING THAT OF ROTTEN EGGS, the acid is *Hydrosulphuric ;* or *Sulphur* combined with a metal (438). (Confirm 439, 443.)

(*c*) If the dilute sulphuric acid gives a PALE YELLOW OR BROWN COLOR to the solution, the acid is probably *Hydriodic*, in which case iodine is set free, and being slightly soluble, colors the liquid (437). (Confirm 434, 435, 436). (See also 559 *b*.)

558. If no effervescence is produced by the dilute sulphuric acid, a portion of the original neutral solution is tested with CHLORIDE OF BARIUM ($BaCl$); if this produces NO PRECIPITATE, the student may pass on to (559); but if a PRECIPITATE APPEARS, the acid is probably either *Sulphuric*, *Phosphoric*, *Boracic*, or *Silicic*,[1]

[1] Arsenious or arsenic acid, if present, would also cause a precipitate with chloride of barium ; but its presence will have been already ascertained during the examination for the base of the salt (544).

since baryta forms with each of them an insoluble salt.

Should any base have been found to be present, which forms an insoluble or sparingly soluble chloride, such as lead, silver, &c. (see Table IX, in the Appendix), a precipitate would here be formed by the chlorine of the chloride of barium. In such cases, a solution of nitrate of baryta may be substituted for the chloride.

(*a*) To distinguish between the acids above enumerated, add a little strong HYDROCHLORIC ACID to the mixture with the precipitate, if the latter DOES NOT DISSOLVE, the acid is probably *Sulphuric*, because the sulphate of baryta is insoluble; while the phosphate, borate, and recently precipitated silicate of baryta, are soluble in hydrochloric acid (403). (Confirm 404, 405.)

If, on the contrary, THE PRECIPITATE DISSOLVES in the hydrochloric acid, the acid is either *Phosphoric*, *Boracic*, or *Silicic*.

(*b*) EVAPORATE A LITTLE OF THE ORIGINAL SOLUTION TO DRYNESS WITH HYDROCHLORIC ACID; treat the residue again with more of the acid; wash the insoluble matter (if any) with water, and examine it before the blowpipe with carbonate of soda; if a TRANSPARENT COLORLESS BEAD is obtained in this way, the acid is probably *Silicic* (427). (Confirm 424.)

(*c*) If it is not silicic, add a little NITRATE OF SILVER to a portion of the original solution; if this gives a PALE YELLOW PRECIPITATE, the acid is probably *Phosphoric* (410). (Confirm 409, 412.)

(*d*) If THE PRECIPITATE THUS PRODUCED IS WHITE, and soluble in nitric acid and in ammonia, it is probably *Boracic* (416). (Confirm 417, 418.)

559. If chloride of barium causes no precipitate, a portion of the original solution must be treated with NITRATE OF SILVER ($AgO,NO_5$): if this causes NO PRECIPITATE, pass on to (560); but if A PRECIPITATE IS PRODUCED, the acid is probably either *Hydrochloric* or *Hydriodic*.

(*a*) If the PRECIPITATE IS WHITE AND CURDY, insoluble in nitric acid, but readily soluble in ammonia, the acid is *Hydrochloric* (429). (Confirm 431.)

(*b*) If THE PRECIPITATE HAS A PALE STRAW COLOR, and is almost insoluble in ammonia, the acid is probably *Hydriodic* (433). (Confirm 435, 436.)

560. If neither chloride of barium nor nitrate of silver give any precipitate, the acid is probably *Nitric* or *Chloric*.

(*a*) WARM A LITTLE OF THE CONCENTRATED SOLUTION WITH STRONG SULPHURIC ACID AND COPPER FILINGS; if ORANGE FUMES are given off, the acid is probably *Nitric* (448). (Confirm 449, 450.)

(*b*) If the acid is not nitric, TEST A SMALL QUANTITY OF THE SOLUTION FOR CHLORIC ACID, in the manner described in paragraphs (454, 455) and confirm (457).

---

## CHAPTER III.

### QUALITATIVE ANALYSIS OF A SIMPLE SALT, CONTAINING ONE BASE AND ONE ACID (OR A SIMPLE METAL), WHICH IS INSOLUBLE OR NEARLY SO IN WATER, BUT SOLUBLE EITHER IN HYDROCHLORIC, NITRIC, OR NITROHYDROCHLORIC ACID (530 ET SEQ.)

### SECTION I.

*Examination for Base.*

561. IN dissolving a substance in acid for the purpose of analysis, it is advisable to avoid using a large excess of the solvent, since it might afterwards interfere with the action of some of the reagents; when a large excess has inadvertently been used, it is consequently necessary to get rid of most of it by evaporation, taking care of course that sufficient acid is left to retain the substance in solution. Most of the substances which are insoluble in water and soluble in acids, owe this solubility to their conversion into compounds which are soluble in water; as when zinc or marble is dissolved in dilute hydrochloric acid, the metallic chloride which is formed ($ZnCl$, or $CaCl$), is soluble in water, and consequently requires no excess of acid to retain it in solution. In some cases, however, when the acid acts merely as a solvent towards the substance, without causing decomposition, it is necessary to have an excess

of acid, to retain it in solution. This is the case with the phosphates of the alkaline earths, and some other salts, which would not dissolve again if the whole of the acid used to dissolve them were to be expelled.

562. When nitric acid has been employed, either alone or in conjunction with hydrochloric acid, it is advisable to expel it, and convert the nitrates into chlorides by adding an excess of hydrochloric acid, filtering if necessary (as when silver, lead, or mercury are present (540)), evaporating the solution nearly to dryness, and adding water or dilute hydrochloric acid. The reason why it is advisable to get rid of the nitric acid is, that it oxidizes and decomposes hydrosulphuric acid, which has to be applied as a test, and thus prevents that reagent playing its proper part in the process.

*Hydrosulphuric Acid Test.*

563. Dilute the acid solution with three or four times its bulk of water,[1] and test a portion of it with HYDROSULPHURIC ACID; if this causes NO PRECIPITATE, pass on to (564); but if A PRECIPITATE IS PRODUCED, refer back to (540 to 546), as this part of the examination is conducted in the same way as when the substance is soluble in water.

*Hydrosulphate of Ammonia Test.*

564. If hydrosulphuric acid gives no precipitate, the base cannot belong to the fourth class. A portion of the solution should next be NEUTRALIZED WITH AMMONIA, if it contains an excess of acid, or if neutral, a little MURIATE OF AMMONIA should be added (547), and subsequently treated with HYDROSULPHATE OF AMMONIA.

If this causes NO PRECIPITATE, pass on to (570); but if A PRECIPITATE IS THROWN DOWN, the base is probably one of those belonging to Class III; or else the precipitate may consist of the *Phosphate* of one of the *Alkaline Earths*, which, in that case, would have been dissolved by the acid, and reprecipitated unchanged when the acid was neutralized by the ammonia and hydrosulphate of ammonia.

[1] If a white precipitate is formed on diluting the acid solution, it is probable that either antimony, bismuth, or tin is present (332, 380, 394).

565. If THE PRECIPITATE CAUSED BY THE HYDROSULPHATE IS BLACK, the base is probably protoxide or peroxide of *Iron*, oxide of *Nickel*, or oxide of *Cobalt*. To distinguish between them, apply the test mentioned in (549).

566. If THE PRECIPITATE THROWN DOWN BY THE HYDROSULPHATE IS FLESH-COLORED, becoming brown by exposure to the air, the base is probably *Protoxide of Manganese* (263). (Confirm 264, 267.)

567. If THE PRECIPITATE IS GREEN, the base is probably *Oxide of Chromium* (247). (Confirm 248, 251, 252.)

568. If THE PRECIPITATE IS WHITE, the base is either *Alumina* or *Oxide of Zinc;* or else the precipitate consists of the phosphate (or ammonio-phosphate) of *Magnesia*, *Lime*, *Baryta*, or *Strontia* (564).[1]

To a portion of the original solution, add POTASH in excess; if THE PRECIPITATE AT FIRST FORMED REDISSOLVES, the base is either *Alumina* or *Oxide of Zinc*, which may be distinguished from each other in the manner described in (551).

569. If THE PRECIPITATE THROWN DOWN BY POTASH IS INSOLUBLE IN EXCESS, it consists probably of an *Earthy Phosphate*, the base being consequently *Magnesia*, *Lime*, *Baryta*, or *Strontia*. In such a case, it is advisable, before proceeding to ascertain which of these is the base present, to separate the phosphoric acid from it. This is done by adding perchloride of iron ($Fe_2Cl_3$) to the acid solution, and subsequently ammonia in slight excess; when the whole of the phosphoric acid is precipitated as perphosphate of iron ($2Fe_2O_3,3HO,3PO_5$), and any excess of perchloride of iron is at the same time precipitated by the ammonia as hydrated peroxide; leaving in solution a chloride of magnesium, calcium, barium, or strontium, together with muriate of ammonia. The solution thus obtained, and filtered from the precipitate of iron, may now be tested with carbo-

[1] Some other salts of the alkaline earths, as the oxalates and borates, would, if present, be thrown down when the solution is neutralized; being, like the phosphates, soluble only in acid solutions. For the sake of simplicity, however, the consideration of such compounds is here omitted.

nate of soda, and further examined according to the directions given in (554).

*Carbonate of Soda Test.*

570. If hydrosulphate of ammonia causes no precipitate, a portion of the original solution is to be tested for the alkaline earths by supersaturating with CARBONATE OF SODA (553, &c.).

571. With regard to the *Alkalies*, it is hardly necessary to allude to them here, as the compounds which they form with all the acids in our list (with the exception of silicic) are soluble in water. In the case of an insoluble alkaline silicate, it is only necessary to evaporate a little of the acid solution of it to dryness, and treat the residue with water. The silicic acid will then be left insoluble (425), and the aqueous solution of the alkaline chloride may be tested for *Potash* and *Soda* in the manner described in paragraph (555).

SECTION II.

*Examination for the Acids.*

572. If the acid is *Arsenious* or *Arsenic*, it will have been detected in the course of the examination for base (563). It is unnecessary to look for *Chloric* acid, since all its salts are soluble in water, and consequently cannot be met with here.

573. A small portion of the substance in the solid state is first treated with HYDROCHLORIC ACID; if this CAUSES EFFERVESCENCE, the acid is probably *Carbonic* (419) (confirm 420); or if the gas which is given off has the SMELL OF HYDROSULPHURIC ACID, the substance under examination is probably a metallic *Sulphide* (438). (Confirm 439, 444.)

574. If the substance is not acted on by the hydrochloric acid, TREAT A LITTLE OF IT WITH NITRIC ACID, and if necessary, boil it.

(*a*) If this CAUSES EFFERVESCENCE, orange fumes of nitrous acid being given off, and sulphur at the same time deposited, the substance is probably a metallic *Sulphide* (439). (Confirm 444.)

(*b*) If the substance DISSOLVES IN NITRIC ACID WITHOUT EFFERVESCENCE, add NITRATE OF SILVER to the acid solution; a WHITE CURDY PRECIPITATE, soluble in ammonia, indicates *Hydrochloric Acid*, the original substance being in that case a *Chloride* (429). (Confirm 431.)

575. Treat a little of the substance in the solid state with STRONG SULPHURIC ACID, and apply heat.

(*a*) If this causes the disengagement of VIOLET VAPOR OF IODINE, the substance under examination is an *Iodide* (436).

(*b*) Add a little ALCOHOL to the acid mixture, which for this experiment should not contain more than a few drops of sulphuric acid, and apply a light to it in a small evaporating dish, placing it in a dark corner, so as to distinguish the color of the flame more readily. If THE FLAME IS GREEN at the edges, the acid is probably *Boracic* (418). (Confirm 417.)

(*c*) Evaporate to dryness a little of the substance after boiling with sulphuric acid, and digest the residue in HOT HYDROCHLORIC ACID; if this leaves a WHITE INSOLUBLE POWDER, which when washed, and heated before the blowpipe with carbonate of soda, fuses into a colorless transparent bead, the acid is *Silicic* (425, 427).

(*d*) Dilute the hydrochloric acid solution formed in (*c*) with water, and add a solution of CHLORIDE OF BARIUM; if this causes A WHITE PRECIPITATE, which is insoluble in nitric acid, the acid is probably *Sulphuric* (403). (Confirm 405, 406).

576. In testing for *Phosphoric Acid*, one of the two following methods may be adopted, according as the base of the salt has been found to belong to the second, third, or fourth class (179).

(*a*) If the base is one of those in Class IV, the diluted acid solution of the substance, containing only a slight excess of acid, is saturated with HYDROSULPHURIC ACID (701); this precipitates the metal, and sets free the phosphoric acid (if present), which remains dissolved in the solution.

$$3CuO,PO_5+3HS=3CuS+3HO,PO_5.$$

The liquid should now be filtered from the precipitated sulphide, concentrated by evaporation, supersaturated with AMMONIA, and tested with SULPHATE OF MAGNESIA; if a white crystalline precipitate is gradually produced, which is insoluble in muriate of ammonia, the acid is probably *Phosphoric* (409). (Confirm 410, 412.)

(*b*) If the base has been found to belong to Class II, III, PERCHLORIDE OF IRON ($Fe_2Cl_3$) is added to a portion of the solution of the substance in hydrochloric acid, and subsequently AMMONIA in slight excess; the phosphoric acid, if present, is precipitated in combination with the iron as perphosphate of iron, together with a little hydrated peroxide of iron, if the perchloride has been added in excess. The precipitate thus formed, containing the whole of the phosphoric acid (if sufficient perchloride of iron has been added), is now well washed with distilled water, and digested with the aid of heat in hydrosulphate of ammonia, by which it is decomposed, sulphide of iron and phosphate of ammonia being formed (413); the latter being soluble, may be separated from the sulphide by filtration, and tested for *Phosphoric Acid* with SULPHATE OF MAGNESIA (409). (Confirm 410, 412.)

577. If the acid is found to be none of those now referred to, it may be nitric, a few of the subnitrates being insoluble in water and soluble in acids. To determine this, a little of the substance in a tube is tested with SULPHURIC ACID AND COPPER FILINGS, when the appearance of ORANGE FUMES will indicate the presence of *Nitric Acid* (448). (Confirm 449, 450.)

---

## CHAPTER IV.

### QUALITATIVE ANALYSIS OF A SIMPLE SALT, CONTAINING ONE BASE AND ONE ACID, WHICH IS INSOLUBLE OR NEARLY SO IN WATER, HYDROCHLORIC, NITRIC, AND NITROHYDROCHLORIC ACIDS (532).

578. IF the salt under examination has been found insoluble in the above solvents, it is probably one of the following substances, viz., a *Silicate* of one of the

metals belonging to Class II, III, or IV; *Sulphate of Lime* ($CaO,SO_3$); *Sulphate of Baryta* ($BaO,SO_3$); *Sulphate of Strontia* ($SrO,SO_3$); *Sulphate of Lead* ($PbO,SO_3$); *Chloride of Lead* ($PbCl$), or *Chloride of Silver* ($AgCl$). Some of these compounds are not altogether insoluble either in water or acids, as the sulphate of lime and chloride of lead; but since they are only very sparingly so, it is possible they may be placed under this head by the experimenter.

579. A small fragment of the substance is moistened with HYDROSULPHATE OF AMMONIA: if it REMAINS WHITE, pass on to (580): but if it BLACKENS, it is probably either *Sulphate of Lead, Chloride of Lead*, or *Chloride of Silver*. A little of the substance in fine powder should in this case be digested for a few hours in HYDROSULPHATE OF AMMONIA, which will gradually decompose it, the metal combining with the sulphur to form an insoluble sulphide, while the acid unites with the ammonia of the hydrosulphate, to form a soluble salt of ammonia. Thus, in the case of sulphate of lead, $2(PbO,SO_3)+NH_4S,HS=2PbS+NH_4O,SO_3+HO,SO_3$.

(*a*) After filtration, the precipitated sulphide is dissolved in NITRIC ACID, and the solution thus obtained may be tested for *Lead* with SULPHURIC ACID (361), and for *Silver* with HYDROCHLORIC ACID (377).

(*b*) The solution filtered from the sulphide is next examined for *Sulphuric Acid* with CHLORIDE OF BARIUM (403), and for *Hydrochloric Acid* (*Chlorine*) with NITRATE OF SILVER (429), confirmatory experiments being made in each case.

580. If the substance REMAINS WHITE WHEN MOISTENED WITH HYDROSULPHATE OF AMMONIA, it is probably either a *Silicate*, or the *Sulphate* of one of the alkaline earths, *Lime, Baryta*, or *Strontia*. A portion of the substance (about twenty to thirty grains) is reduced to fine powder, and intimately mixed with four or five times its weight of dry carbonate of soda. The mixture is placed in a platinum (or porcelain) crucible,[1] and heated to redness,

[1] Great care is necessary in using a platinum crucible, that nothing is heated in it which is likely to corrode it. Compounds of the easily reduced and fusible metals, as tin, antimony, lead, bismuth, &c., substances containing sulphur, as metallic sulphides, caustic alkalies, nitro-hydrochloric acid, besides many other substances, are all more or less

either in a furnace or over a lamp, for about an hour (648). The fused mass, when cool, is digested in dilute hydrochloric acid until it is for the most part dissolved, and a little of the solution is tested for *Sulphuric Acid* with CHLORIDE OF BARIUM (403).

(*a*) If THIS INDICATES THE PRESENCE OF SULPHURIC ACID, the substance is probably the *Sulphate of Lime*, *Baryta*, or *Strontia*, and the acid solution may be neutralized with AMMONIA, and examined for those bases according to the directions given in (554).

(*b*) If NO SULPHURIC ACID IS PRESENT, the substance is probably a *Silicate*. In this case the hydrochloric acid solution, together with any portion that may have resisted solution, is evaporated to dryness, and the residue treated with hydrochloric acid, and subsequently with water; if a WHITE INSOLUBLE POWDER remains, which fuses with carbonate of soda before the blowpipe, into a clear colorless bead, *Silicic Acid* is present (425, 427).

The solution obtained in (*b*), by treating the dry residue with hydrochloric acid and water, contains the base with which the silicic acid was combined; and may be examined according to the directions given in paragraphs 563 et seq.

---

# CHAPTER V.

## QUALITATIVE ANALYSIS OF A MIXTURE OF TWO OR MORE SALTS WHICH MAY CONTAIN ALL THE BASES AND ACIDS IN THE LIST (179).

### *Introductory Remarks.*

581. UNLESS we have reason to know that a substance intended for analysis contains only one base and one acid, it is necessary to assume that it *may* contain any or all of the more common saline compounds. Such an analysis is of course considerably more complicated

injurious. When a platinum crucible is heated in a furnace or open fire, it must be placed in a covered earthen crucible to protect it from injury; a little pounded magnesia should be interposed between them, to prevent their sticking together, as at a high temperature the surface of the earthenware is liable to fuse (648).

than that of a single salt; and consequently the necessity of having a well-devised scheme of experiments is here even greater than in the former case, when only base and acid had to be determined.

The method of dealing with such a mixture is, first to separate the whole of the metals of the fourth class (if any are present) by passing hydrosulphuric acid gas through a solution of the substance acidified with hydrochloric acid, and filtering the solution from the precipitate: the precipitate is then dissolved in acid, and tested successively for each of the metals of the fourth class. The filtered solution, containing all the bases but those of the fourth class, is then neutralized, and treated with hydrosulphate of ammonia, which throws down all the metals of the third class (if any are present); and the precipitate filtered from the solution is dissolved in acid, and tested successively for each metal of the third class. The solution filtered from the sulphides can now only contain any metals of the first and second class that may be present, which may readily be distinguished by a few simple tests.

582. The student must be careful when making these experiments, that he adds sufficient of the various reagents, to throw down *the whole* of the metals affected by them, since any traces of the metals belonging to a class supposed to have been entirely removed from the solution, would materially interfere with the indications afforded by the subsequent tests. For example, in the analysis of a mixture of a salt of lead and a salt of lime, if sufficient hydrosulphuric acid were not passed through the solution to separate the whole of the lead, a black precipitate of sulphide of lead would be formed on the addition of hydrosulphate of ammonia to the filtered liquid, indicating the presence of one or more metals of the third class, none of which are really present. On the other hand, the addition of a large excess of any of the reagents is also to be avoided, as being not only useless and wasteful, but in many cases mischievous.

Both these errors may be avoided by adding the reagents in small successive portions; and when the experimenter has reason to think that he has added sufficient, let him filter a few drops of the mixture, and apply to the solution a little more of the reagent: if this produces no further precipitate, he may conclude that enough has been added.

583. When a class of metals has been precipitated by either of the general reagents mentioned in (581), it is always advisable, before proceeding to apply any of the subsequent tests to the filtered solution, to ascertain whether it contains any other fixed bases; as if it does not, the examination of it need not be proceeded with. This is readily known by evaporating a drop or two of the solution on platinum foil, and heating it to redness; when, if no residue is left, it may safely be concluded that all the bases (except ammonia, which must be looked for in a separate portion (602) ) have been already separated. In the course of an analysis, especially of a complicated substance, it is often necessary to have several solutions in hand at the same time; to avoid confusion, each of these should be labelled with a bit of gummed paper, with a letter or mark upon it, referring to a corresponding letter in the note book (7).

584. As one portion of the substance to be analyzed has to be carried through several operations, it is advisable that the quantity operated on should not be very small. When the substance is a solid, twenty or thirty grains may be used; and when in solution, an ounce or two (according to the degree of concentration) will be found a convenient quantity.

---

## CHAPTER VI.

### QUALITATIVE ANALYSIS OF A MIXTURE OF SALTS WHICH MAY CONTAIN ALL THE BASES AND INORGANIC ACIDS IN THE LIST (179), AND WHICH IS READILY SOLUBLE IN WATER (529).[1]

### SECTION I.

*Examination for Bases.*

585. The solution is first rendered slightly acid by

[1] It is of course impossible that such a solution can exist, containing *all* the bases and acids in the list, since several of them would form

the addition of a few drops of HYDROCHLORIC ACID: if this causes NO PRECIPITATE, pass on to (586); but if A WHITE PRECIPITATE IS PRODUCED, it is owing to the presence of *Silver*, *Lead*, or *Protoxide of Mercury*. In this case, add HYDROCHLORIC ACID as long as it causes any precipitate; filter the liquid, and wash the insoluble chloride.

Place a small portion of the moist precipitate in a test-tube, and treat it with AMMONIA.

(*a*) If THE PRECIPITATE DISSOLVES COMPLETELY, it consists wholly of chloride of silver, proving of course the presence of *Silver* in the substance under examination (377). (Confirm 374, 378.)

(*b*) If THE PRECIPITATE IS BLACKENED, and not wholly dissolved, by the ammonia, it probably contains protochloride of *Mercury* (338). (Confirm 336, 344.)

(*c*) If it appears to be UNAFFECTED BY THE AMMONIA, it is chloride of *Lead* (362). (Confirm 361, 363, 366.)

(*d*) If IT DOES NOT WHOLLY DISSOLVE (*b* and *c*), pass the ammoniacal mixture through a filter, and neutralize the solution with NITRIC ACID: if *Silver*, in addition to lead or mercury, is present, it will be reprecipitated as chloride (377). (Confirm 374, 378.)

*Hydrosulphuric Acid Test.*

586. The solution, acidified with hydrochloric acid, and filtered if necessary from the precipitate, is now treated with HYDROSULPHURIC ACID gas, which must be passed through it until, after removing the delivering tube, and blowing the air from the surface of the solution, the latter smells distinctly of the gas. If NO PRECIPITATE IS PRODUCED, even on boiling the mixture, pass on to (593); but if on the contrary, A PRECIPITATE FALLS, one or more of the metals of the fourth class are present: if this is the case, the precipitate must be separated from the solution by filtration, and washed

salts which are insoluble in water, as baryta and sulphuric acid, oxide of silver and hydrochloric acid, &c. It is consequently unnecessary, after having determined the bases in a mixture of salts soluble in water, to look for any acids which form with them salts that are insoluble. (See Table of Solubilities in the Appendix.)

with distilled water, until a drop of the washings leaves no fixed residue when evaporated on platinum foil, the filtered solution being carefully reserved for further examination (593).[1]

587. If THE PRECIPITATE PRODUCED BY HYDROSULPHURIC ACID IS YELLOW, it may be owing to the presence of *Arsenic* (307), or *Peroxide of Tin* (388). In this case it is advisable first to dry a little of the precipitate, and to test it for arsenic with black flux (303). If arsenic is thus found to be present, we may at once conclude that no bases, with the exception of the alkalies, can be present, because the compounds of arsenious and arsenic acid, with all the other bases, are more or less insoluble in water, and consequently cannot exist in an aqueous solution, like that now under consideration. When therefore arsenic is found, the student may at once pass on to (601): and having concentrated the solution, and divided it into three portions, proceed to examine it for potash, soda, and ammonia.

588. When arsenic is not present, or WHEN THE PRECIPITATE CAUSED BY HYDROSULPHURIC ACID IS ANY OTHER COLOR THAN YELLOW, it must, after being well washed (586), be separated from the filter, and digested, with the aid of a gentle heat, for about a quarter of an hour, in a small basin, with HYDROSULPHATE OF AMMONIA. If the sulphides of *Antimony* or of *Tin* are present, they will dissolve in the hydrosulphate, forming soluble double sulphides, while the sulphides of the other metals of the fourth class that may be present, will remain undissolved.[2]

If the sulphides, or any portion of them REMAIN UNDISSOLVED BY THE HYDROSULPHATE, the mixture must

[1] In *qualitative* analysis, the first portions only of the washings need be retained (unless we possess only a small quantity of the substance), as the rest would only uselessly dilute our solution; but in *quantitative* analysis, it is necessary to retain the whole of them, as their rejection would occasion a serious deficiency in the weight of the substance under examination.

[2] If copper is present, which may be readily ascertained by adding ammonia in excess to the original solution (369), sulphide of potassium must be substituted for the hydrosulphate of ammonia, because the latter would dissolve some of the sulphide of copper.

be filtered, and the insoluble portion well washed; the solution will then have to be examined for *Antimony* and *Tin*, and the insoluble portion for *Lead*, *Bismuth*, *Copper*, and *Mercury*, thus :—

589. Dilute the hydrosulphate of ammonia solution with about an equal bulk of water, and supersaturate it with ACETIC ACID, which will cause a precipitation of sulphur (440) and of the sulphides of tin and antimony, if they are present (330). This precipitate is washed with water, dried, and a little of it gently ignited on platinum foil, to prove whether it contains anything more than sulphur, in which case, a fixed residue is left; while, if THE WHOLE VOLATILIZES, the examination of the matter precipitated by the acetic acid need not be proceeded with, and the student may pass on to (590), neither tin nor antimony being present.

If, on the other hand, A RESIDUE IS LEFT ON THE PLATINUM FOIL after ignition, either *Tin* or *Antimony* (or both) are present; in this case the precipitate may be boiled for about half an hour in a test-tube with strong HYDROCHLORIC ACID, and, after standing for a short time to allow the undissolved sulphur to subside, the clear solution, which may contain the chloride of antimony and perchloride of tin,[1] is poured off.

(*a*) Dilute a portion of the hydrochloric acid solution with four or five times its bulk of WATER: if IT BECOMES MILKY, *Antimony* is probably present (332). (Confirm 333, 334.)

(*b*) Evaporate another portion of the hydrochloric acid solution to dryness, mix the residue with carbonate of soda, and heat it in the inner flame of the blowpipe; if MALLEABLE METALLIC GLOBULES are thus formed, *Tin* is probably present (379). (Confirm 384, 386.)

(*c*) To ascertain whether the tin existed as protoxide or peroxide, a little of the original solution may be tested with TERCHLORIDE OF GOLD, which gives a PURPLE PRECIPITATE with *Protosalts of Tin* (386).

[1] If the tin existed as protoxide, and consequently as protosulphide (SnS) in the precipitate thrown down by hydrosulphuric acid, it will have been converted into the persulphide ($Sns_2$) by the action of the excess of sulphur usually present in the hydrosulphate of ammonia (382.)

590. The portion of the sulphides which did not dissolve in the hydrosulphate of ammonia (588), must now be examined for *Lead*, *Bismuth*, *Copper*, and *Mercury*.

The precipitate is removed from the filter, into a small evaporating basin, and boiled with STRONG NITRIC ACID for about a quarter of an hour: the solution is then diluted with water, and if anything remains undissolved, filtered.

591. The undissolved matter may contain sulphide of mercury, sulphur and sulphate of lead; the sulphuric acid of which will have been formed by the action of the nitric acid on the sulphur of the sulphides.

(*a*) Heat a little of it on platinum foil; if a WHITE RESIDUE is left after ignition, which blackens when moistened with hydrosulphate of ammonia, *Lead* is probably present (361). (Confirm 366.)

(*b*) Mix another portion of the dried residue with carbonate of soda, and heat it in a hard glass tube; if *Mercury* is present, METALLIC GLOBULES will condense in the upper part of the tube (336).

592. The nitric acid solution (590) may contain lead, copper, and bismuth.

(*a*) EVAPORATE THE SOLUTION NEARLY TO DRYNESS, AND DILUTE IT WITH WATER; if a WHITE PRECIPITATE is produced, *Bismuth* is probably present (394). (Confirm, 395, 397.)

(*b*) To the solution formed in (*a*), filtered, if necessary, from the precipitate, add DILUTE SULPHURIC ACID; if this causes a WHITE PRECIPITATE, *Lead* is probably present (361). (Confirm 363, 366).

(*c*) To another portion of the clear solution (*a*) add AMMONIA in slight excess; if this gives a PALE BLUE PRECIPITATE, which readily redissolves in excess of ammonia, forming a BLUE SOLUTION, *Copper* is present (369).

*Hydrosulphate of Ammonia Test.*

593. A few drops of the solution filtered from the precipitate thrown down by hydrosulphuric acid, or which failed to produce a precipitate with it (586), are now evaporated on platinum foil, to ascertain whether it

contain any other fixed base; and if it is found to leave NO RESIDUE the examination need not be proceeded with; but if A RESIDUE IS LEFT, a small portion of the solution is neutralized with AMMONIA in a test-tube, and treated with HYDROSULPHATE OF AMMONIA. If this gives NO PRECIPITATE, the solution does not contain any of the metals in the third class, and the student may pass on to (596); but if A PRECIPITATE APPEARS, the whole of the liquid is similarly treated, first with AMMONIA and then with the HYDROSULPHATE; a little MURIATE OF AMMONIA being also added, unless the solution contained a decided excess of hydrochloric acid, in which case, the muriate would be formed on neutralizing the acid with ammonia.[1]

When the hydrosulphate has been added as long as it causes any precipitate, the liquid is filtered, and the precipitate well washed, until a drop of the washings, when evaporated on platinum foil and ignited, leaves no fixed residue,[2] the clear solution being retained for further examination (596).

594. The precipitate is dissolved in NITROHYDROCHLORIC ACID, heat being applied if necessary; and if any sulphur remains undissolved, the mixture is filtered. The solution thus obtained may contain *Peroxide of Iron*, *Alumina*, and the oxides of *Chromium*, *Manganese*, *Zinc*, *Nickel*, and *Cobalt*.

(*a*) AMMONIA is now added in excess, which precipitates the peroxide of iron, alumina, and oxide of chromium, while the four remaining oxides, if present, are redissolved. If NO PRECIPITATE REMAINS, pass on to (595).

(*b*) If A PRECIPITATE IS FORMED BY THE AMMONIA, the mixture is filtered; and the precipitate, after being washed, is redissolved in HYDROCHLORIC ACID.

(*c*) POTASH is added in excess to the hydrochloric acid solution: if this causes A PRECIPITATE WHICH IS INSOLUBLE IN EXCESS, *Peroxide of Iron* is probably present (281). (Confirm 282.)

(*d*) The potash solution, filtered if necessary, from the precipitate (*c*), may contain alumina and oxide of

[1] See note to 547. [2] See note to 586.

chromium. If oxide of chromium is present, the solution will probably have a GREEN COLOR, and on boiling the potash solution, the hydrated oxide of *Chromium* gradually separates as a dark green precipitate, leaving the solution colorless (249). (Confirm 251, 252.)

(*e*) If *Alumina* is present, it will be PRECIPITATED FROM THE POTASH SOLUTION on the addition of MURIATE OF AMMONIA (242), especially if the excess of alkali be first nearly neutralized with hydrochloric acid. (Confirm 240, 245.)

595. The ammoniacal solution (594 *a*) is now to be examined. A drop or two are first evaporated on platinum foil, when if NO FIXED RESIDUE remains, proving the absence of fixed bases, the examination need not be proceeded with; but if any RESIDUE IS LEFT, the ammoniacal solution is treated with HYDROSULPHATE OF AMMONIA as long as it produces any precipitate. This precipitate is redissolved in nitrohydrochloric acid, and the solution supersaturated with POTASH. If a PRECIPITATE IS FORMED, the mixture is filtered, and the precipitate washed.

(*a*) The filtered solution may contain oxide of *Zinc*. If this is the case, the addition of HYDROSULPHURIC ACID to the potash solution throws down the WHITE SULPHIDE (256). (Confirm 260, 261.)

(*b*) The precipitate (if any) thrown down by potash, which may contain the oxides of manganese, cobalt, and nickel, is warmed with a solution containing ammonia and carbonate of ammonia: if any of the PRECIPITATE REMAINS UNDISSOLVED, oxide of *Manganese* is probably present. (Confirm 267, 268.)

(*c*) If any fixed residue is left when a drop of the ammoniacal liquid formed in (*b*) is evaporated on platinum foil, it may contain the oxides of cobalt and nickel. In this case, evaporate the solution to dryness, and test a little of the residue with borax before the blowpipe, for *Cobalt* (299). (Confirm 295, 296.)

(*d*) Add a little hydrochloric acid to the other portion of the residue formed in (*c*), and expel the greater part of it by evaporation, leaving only a slight excess of acid. Dissolve it in water, and add a solution of

CYANIDE OF POTASSIUM until any precipitate that may be formed is entirely redissolved: if the addition of dilute sulphuric acid to the solution gradually causes a precipitate, *Nickel* is probably present (291). (Confirm 287, 288.)

*Carbonate of Ammonia Test.*

596. The solution filtered from the precipitate thrown down by hydrosulphate of ammonia, or which failed to produce a precipitate with that reagent, is now to be tested. A few drops are first evaporated on platinum foil, and if NO FIXED RESIDUE remains, it need not be examined for any other fixed basis, and ammonia only will have to be looked for in addition to the bases already discovered (see 602 for the method of testing for ammonia).

If, on the contrary, a RESIDUE IS LEFT, the solution is boiled for some time to expel the hydrosulphuric acid, a little hydrochloric acid having previously been added if hydrosulphate of ammonia had been used.

$$NH_4S,HS+HCl=NH_4Cl+2HS.$$

If any sulphur is precipitated in this process, it must be separated by filtration.

The solution is now mixed with a little muriate of ammonia, unless already formed by neutralizing an excess of ammonia or the hydrosulphate with hydrochloric acid; CARBONATE OF AMMONIA, mixed with a little ammonia (712), is added as long as it causes any precipitate, and the solution is boiled. If NO PRECIPITATE is thrown down, the student may pass on to (598), neither lime, baryta, nor strontia being present; but if A PRECIPITATE FALLS, it is owing to the presence of one or more of those bases. In this case the mixture is filtered, and the precipitate (which may contain the carbonates of the alkaline earths just mentioned) is well washed, the filtered solution being retained for subsequent examination (598).

597. The precipitate is dissolved in a small quantity of HYDROCHLORIC ACID; the solution thus formed is neutralized with AMMONIA, and divided into three portions.

(*a*) Add to the first, SULPHATE OF SODA as long as it causes any precipitate, and filter. If OXALATE OF AMMONIA gives with the filtered solution a WHITE PRECIPITATE, *Lime* is present (216). (Confirm 219.)

(*b*) To the second portion add a solution of SULPHATE OF LIME; if this causes an IMMEDIATE WHITE PRECIPITATE, *Baryta* is probably present (225). (Confirm 228.)

(*c*) The third portion is evaporated to dryness, and a little of the residue heated before the blowpipe; if the FLAME IS TINGED WITH A CARMINE COLOR, *Strontia* is probably present (236). (Confirm 232, 233.)

598. The liquid filtered from the precipitate caused by the carbonate of ammonia, or which failed to give a precipitate with that reagent (596), is now to be examined. If it LEAVES ANY RESIDUE when evaporated on platinum foil, it may contain *Magnesia*, *Potash*, and *Soda*.

When lime, baryta, or strontia have been detected in the mixture, it is always advisable to test a little of the solution filtered from the carbonates, with OXALATE OF AMMONIA and SULPHATE OF SODA, in order to see whether the whole of the three earths had been separated by the carbonate of ammonia: if either of the tests shows traces of them, the solution is to be again boiled with a fresh addition of ammonia and carbonate of ammonia, until the whole of them is removed.

599. A little of the ammoniacal solution, moderately concentrated, is now tested with PHOSPHATE OF SODA; if this causes a WHITE CRYSTALLINE PRECIPITATE, *Magnesia* is present (206). (Confirm 208, 209.)

600. If MAGNESIA IS NOT PRESENT, the ammoniacal solution is evaporated to dryness, and the residue ignited to expel the ammoniacal salts; the residue is then dissolved in as small a quantity as possible of water, and the solution divided into three portions, to be tested according to the directions given in (601).

When MAGNESIA IS PRESENT, it is necessary to separate it from the solution by some reagent which does not contain soda, since that alkali has still to be sought for in the solution. In such a case, the following is the best method. The remaining portion of the ammoniacal solution is evaporated to dryness, and the residue ignited in a small platinum crucible, to expel the ammoniacal salts. The fixed matter is dissolved in a little water, treated with a saturated SOLUTION OF CAUSTIC BARYTA, and allowed to stand some little time, to cause the whole of the magnesia to precipitate (208). The mixture is then filtered; DILUTE SULPHURIC ACID is added in very slight excess to the clear solution, to throw down the whole of the baryta; and the liquid, after boiling, is filtered. The filtered solution is evaporated to dryness, to expel the excess of sulphuric acid; and the residue is gently ignited; this is redissolved in the smallest possible quantity of water, and the solution divided into three portions. If NO RESIDUE IS LEFT after the ignition, neither of the fixed alkalies is present.

601. (*a*) The first portion is tested with BICHLORIDE OF PLATINUM, for *Potash* (185).

(*b*) The second portion is acidified with TARTARIC ACID, also for *Potash* (186).

(*c*) The third portion is tested with ANTIMONIATE OF POTASH, for *Soda* (189). (Confirm 188, 190.)

602. As the substance under examination has to be ignited during the analysis, it is of course impossible that ammonia can be detected with the other alkalies in the process now described. A portion of the original solution is therefore to be mixed with an excess of caustic POTASH and warmed; if *Ammonia* is present, it may be detected by the SMELL, or by holding a rod moistened with HYDROCHLORIC ACID near the mouth of the test-tube (195).

## SECTION II.

### *Examination for the Acids.*

603. The original solution of the substance is first examined with litmus and turmeric paper; if it has an acid reaction, a little of it is tested in the manner described in (535 *b*), and if it is found to owe its acid reaction to the presence of free acid, the solution must be carefully neutralized with dilute POTASH; but if it is only a metallic salt which caused it, the solution may

be considered neutral. If, on the other hand, the liquid has an alkaline reaction, which may be be owing to the presence either of a free alkali or of alkaline carbonates or hydrosulphates, it must be rendered perfectly neutral by HYDROCHLORIC ACID, and boiled to expel carbonic or hydrosulphuric acids, if either are present.

604. To a small portion of the original solution add HYDROCHLORIC ACID in excess: if this causes EFFERVESCENCE, carbonic and hydrosulphuric acids may be present. If NO EFFERVESCENCE takes place, pass on to (605).

(*a*) If the GAS IS INODOROUS, or when passed into lime-water causes a white precipitate, *Carbonic acid* is present (419, 420).

(*b*) If the gas has a DISAGREEABLE SMELL, and when passed into a solution of acetate of lead, causes a black or brown precipitate, *Hydrosulphuric Acid* (*Sulphur*) is probably present (438). (Confirm 439, 444.)

*Nitrate of Baryta Test.*

605. Add to the original solution of the substance, neutralized if necessary (603), NITRATE OF BARYTA as long as it causes any precipitate. If NO PRECIPITATE is formed, pass on to (607). The mixture is filtered, and the precipitate washed, the clear solution being reserved for further examination (607). The precipitate may contain *Sulphuric*, *Arsenic*, *Arsenious*,[1] *Phosphoric*, *Boracic*, and *Silicic Acids*, in combination with baryta.

(*a*) The precipitate is heated with strong hydrochloric acid, and the mixture evaporated to dryness. The residue is again warmed with hydrochloric acid, and the liquid, after boiling, is diluted with a little water, and filtered if ANYTHING REMAINS UNDISSOLVED, in which case sulphuric and silicic acids may be present.

(*b*) Add an excess of HYDROCHLORIC OR NITRIC ACID to a small portion of the original solution, and then a few drops of NITRATE OF BARYTA: if this causes a WHITE PRECIPITATE, insoluble when the mixture is heated, *Sulphuric Acid* is present (403). (Confirm 405, 406.)

[1] If arsenious or arsenic acids are present, they will have been already detected in the course of the examination for bases (587).

(c) To another small portion of the original solution add HYDROCHLORIC ACID, and evaporate the solution to dryness; if any of the dry RESIDUE IS INSOLUBLE IN HYDROCHLORIC ACID, *Silicic Acid* is probably present (425). (Confirm 427.)

606. The hydrochloric acid solution (605 *a*) may contain phosphoric and boracic (as well as arsenious and arsenic) acids, the compounds of those acids with baryta being soluble in hydrochloric acid.

(a) Test a little of the original neutral solution with MURIATE OF AMMONIA and SULPHATE OF MAGNESIA;[1] if a CRYSTALLINE PRECIPITATE is formed, either immediately or after standing a short time, *Phosphoric Acid* is probably present (409). (Confirm 410, 412.)

(b) Add a little SULPHURIC ACID to a small portion of the original solution, or of the substance in the solid state, and evaporate the mixture to dryness. Treat the residue with ALCOHOL, and after allowing it to digest a short time, set fire to it in a dark place: if the FLAME OF THE ALCOHOL IS COLORED GREEN, *Boracic Acid* is probably present (418). (Confirm 416, 417.)

*Nitrate of Silver Test.*

607. The solution filtered from the precipitate caused by nitrate of baryta, or in which that reagent failed to produce a precipitate (605), is now examined. It may contain *Hydrochloric*,[2] *Hydriodic*, *Nitric*, and *Chloric Acids:* and in addition to these, in case the original solution was dilute, or contained ammoniacal salts, traces of boracic, arsenious, and arsenic acids. The solution is treated with NITRATE OF SILVER; if any PRECIPITATE IS PRODUCED, the mixture is filtered, and the clear solution reserved for subsequent examination (608).

(a) Add an excess of AMMONIA to the precipitate; if

[1] If the substance has been found to contain any base that causes a precipitate with sulphuric acid (see Table of Solubilities in the Appendix), chloride of magnesium must be used instead of the sulphate.

[2] If hydrochloric acid has been used to neutralize the solution (603), it will of course be precipitated here, so that it will be necessary to test a little of the original solution, neutralized with *nitric acid*, to ascertain whether any hydrochloric acid or chlorine is present in it.

IT DOES NOT WHOLLY DISSOLVE, *Hydriodic Acid* (*Iodine*), is probably present (433). (Confirm 435, 436.)

(*b*) The ammoniacal solution formed in (*a*) is supersaturated with NITRIC ACID: if a WHITE CURDY PRECIPITATE is thrown down, *Hydrochloric Acid* (*Chlorine*) is probably present (429). (Confirm 430, 431.)

(*c*) The acid solution formed in (*b*) may contain traces of arsenious, arsenic, and boracic acids. The first two will, if present, have been already found in the examination for the bases; the latter may be detected in the manner described in (606 *b*).

608. The solution filtered from the precipitate thrown down by nitrate of silver, or in which that reagent failed to produce a precipitate (607), may contain *Nitric* and *Chloric Acids;* but as nitric acid has been added to it in the nitrates of baryta and silver, some of the original solution must be used in this part of the examination.

(*a*) To a small portion of the original solution, add SULPHURIC ACID AND COPPER FILINGS; if ORANGE FUMES are disengaged, *Nitric Acid* is probably present (448). (Confirm 449, 450.)

(*b*) Moisten a portion of the original substance in the dry state, with STRONG SULPHURIC ACID: if a GREENISH GAS is evolved, *Chloric Acid* is probably present (457). (Confirm 454, 455, 456.)

---

## CHAPTER VII.

### QUALITATIVE ANALYSIS OF A MIXTURE OF TWO OR MORE SALTS, WHICH MAY CONTAIN ALL THE BASES AND INORGANIC ACIDS IN THE LIST (179), AND WHICH IS INSOLUBLE, OR NEARLY SO, IN WATER, BUT READILY SOLUBLE IN HYDROCHLORIC, NITRIC, OR NITROHYDROCHLORIC ACID (530).

#### SECTION I.

*Examination for the Bases.*

609. When it has been found necessary to use a large excess of acid to dissolve the substance, it is advisable,

before beginning the analysis, to expel the greater part of it by evaporation (561).

610. When the substance has been dissolved in hydrochloric acid or nitrohydrochloric acid, it is unnecessary to examine it for silver or protoxide of mercury, because the corresponding chlorides of those metals are insoluble in hydrochloric acid. When the solution has been made in nitric acid, a little HYDROCHLORIC ACID is first added to the solution: if this causes NO PRECIPITATE, pass on to (611); but if A WHITE PRECIPITATE IS PRODUCED, *Silver*, *Protoxide of Mercury*, and *Lead*, may be present. In this case add HYDROCHLORIC ACID as long as it causes any further precipitate, filter, and examine the precipitate with AMMONIA, as already described (585), the solution being retained for further examination (611).

611. When nitric acid has been used in dissolving the substance, either alone or in conjunction with hydrochloric acid, it is advisable to expel it before proceeding to test the solution with hydrosulphuric acid; because when nitric acid is present in a solution, the hydrosulphuric acid is oxidized by it, and is thus prevented from acting in the usual manner on the metallic oxides present. The solution containing nitric acid should therefore be mixed with an excess of hydrochloric acid, filtered if necessary, and evaporated nearly to dryness. The concentrated solution is then diluted with water, and if any MILKINESS IS PRODUCED on dilution, owing to the presence of antimony, bismuth, or tin, it may be disregarded, as it will not interfere with the action of the hydrosulphuric acid.

*Hydrosulphuric Acid Test.*

612. HYDROSULPHURIC ACID GAS is now passed through the dilute acid solution, until it is saturated; if this causes NO PRECIPITATE, even when the mixture is boiled, pass on to (614); but if A PRECIPITATE IS PRODUCED, it is owing to the presence of one or more metals of the fourth class. The precipitate is to be separated by filtration, and washed as long as a drop of the washings leaves any fixed residue when evaporated on platinum

foil; the clear liquid being retained for subsequent examination (614).

613. The precipitate, which may contain the sulphides of all the metals in the fourth class, after being well washed, is digested with HYDROSULPHATE OF AMMONIA; and the portions, both soluble and insoluble in the hydrosulphate, are examined in the manner described in (588 to 592).

As, however, in this case, arsenic may coexist with any of the other bases of the fourth class (such compounds being for the most part soluble in acids), it is necessary to examine the precipitate thrown down by acetic acid from the solution of the sulphides in hydrosulphate of ammonia, FOR ARSENIC, as well as for antimony and tin. This may be easily done by applying the reduction test (303).

*Hydrosulphate of Ammonia Test.*

614. The solution filtered from the precipitate thrown down by hydrosulphuric acid, or which failed to produce a precipitate with it (612), is now treated with AMMONIA and HYDROSULPHATE OF AMMONIA, and examined according to the directions given in (593 to 595). As, however, a mixture such as we are now considering, which is insoluble in water, may contain the *Earthy Phosphates*, those compounds, if present, will be thrown down by the ammonia and hydrosulphate (564); and it is necessary to examine the precipitate for lime, magnesia, baryta, and strontia, in addition to the metals belonging to Class III. These earthy phosphates, if present, will be thrown down by potash, together with any iron that may be present (594, *c*): that precipitate may consequently contain peroxide of iron, together with the *Phosphate of Lime*, *Magnesia*, *Baryta*, and *Strontia*.

(*a*) The precipitate (594, *c*) is dissolved in hydrochloric acid, and to a small portion of the solution thus formed, FERROCYANIDE OF POTASSIUM is added; if this causes a BLUE PRECIPITATE, *Peroxide of Iron* is present (282).

(*b*) To the rest of the solution, PERCHLORIDE OF IRON is added, and afterwards an excess of AMMONIA; this throws down the whole of the iron as hydrated peroxide, which carries with it in combination any phosphoric acid that may be present; while the solution contains, in the form of chlorides, the *Alkaline Earths* which were previously combined with *Phosphoric Acid* (569).

(*c*) The liquid thus obtained is tested, after filtration, for fixed bases, by evaporating a drop on platinum foil; and if A RESIDUE IS LEFT, the solution is examined for *Lime*, *Baryta*, *Strontia*, and *Magnesia*, CARBONATE and MURIATE of AMMONIA being added, and the precipitate and solution treated in the manner described in (597 to 599).

*Carbonate of Ammonia Test.*

615. The solution filtered from the precipitate caused by hydrosulphate of ammonia, or in which that reagent failed to produce any precipitate, is now examined for the *Alkaline Earths* and *Alkalies*, in the manner already described in the case of substances which are soluble in water (596 to 602).

## SECTION II.

*Examination for the Acids.*

616. A little of the substance under examination is mixed with strong hydrochloric acid: if EFFERVESCENCE TAKES PLACE, carbonic and hydrosulphuric acids may be present.

(*a*) If the gas which is evolved causes a WHITE PRECIPITATE when passed into LIME-WATER, *Carbonic Acid* is present (419, 420).

(*b*) If the gas causes a BLACK OR BROWN PRECIPITATE, when passed into a solution of ACETATE OF LEAD, *Hydrosulphuric Acid* (*Sulphur in a Sulphide*) is present (438). (Confirm 444.)

617. The solution of the substance in hydrochloric acid is now examined for sulphuric, phosphoric, and silicic acids.

(*a*) A portion of the hydrochloric acid solution is tested with CHLORIDE OF BARIUM; if this causes a WHITE PRECIPITATE, which is insoluble when warmed with an excess of hydrochloric acid, *Sulphuric Acid* is present (403). (Confirm 405, 406.)

(*b*) A little of the hydrochloric acid solution is evaporated to dryness, and the residue treated with HYDROCHLORIC ACID; if a WHITE INSOLUBLE POWDER IS LEFT, which, when washed, and heated before the blowpipe with carbonate of soda, fuses into a transparent colorless bead, *Silicic Acid* is present (425, 427).

(*c*) Phosphoric acid may be detected in the following manner. To a portion of the hydrochloric solution, PERCHLORIDE OF IRON is added, and then AMMONIA in slight excess; the precipitate thus produced is well washed on a filter, digested, with the aid of a gentle heat, in HYDROSULPHATE OF AMMONIA, and filtered. If the solution thus obtained gradually throws down, when concentrated, a WHITE CRYSTALLINE PRECIPITATE with SULPHATE OF MAGNESIA, *Phosphoric Acid* is probably present (409, 413). (Confirm 410, 412.)

618. A portion of the substance is treated with STRONG NITRIC ACID, and, if necessary, warmed.

(*a*) If ORANGE FUMES ARE EVOLVED, and a pale yellow deposit of sulphur is produced, a metallic *Sulphide* is present (439). (Confirm 444.)

(*b*) Add to the nitric acid solution a few drops of NITRATE OF SILVER; if this CAUSES A PRECIPITATE, wash it on a filter, and digest in AMMONIA. If a WHITE CURDY PRECIPITATE is thrown down when the ammoniacal solution is neutralized with NITRIC ACID, *Hydrochloric Acid* (*a Metallic Chloride*) is present (429). (Confirm 430, 431.)

619. Test a little of the substance for *Boracic Acid* in the manner described in (606, *b*).

620. If the substance disengages VIOLET-COLORED FUMES, when warmed with STRONG SULPHURIC ACID, *Iodine* (*a Metallic Iodide*) is present (436).

621. Place a fragment of the dry substance on ignited charcoal: if this occasions DEFLAGRATION, *Nitric Acid* is probably present (447). (Confirm 448, 450.)

622. *Chloric Acid* need not be looked for in compounds which are insoluble in water, since all the chlorates are readily soluble.

## CHAPTER VIII.

### QUALITATIVE ANALYSIS OF A MIXTURE OF TWO OR MORE SALTS, WHICH MAY CONTAIN ALL THE BASES AND INORGANIC ACIDS IN THE LIST (179), AND WHICH IS INSOLUBLE, OR NEARLY SO, IN WATER AND ACIDS.

623. THE compounds most likely to be found in such a mixture, are those enumerated in (578). The best method of rendering such a substance soluble, is to fuse it with CARBONATE OF SODA (580), either in a platinum or porcelain crucible. If any metals of the fourth class are present, which may generally be ascertained by moistening a small fragment of the substance with HYDROSULPHATE OF AMMONIA (579), it is safer to use a porcelain crucible.[1] When this is done, a little silica and alumina will generally be dissolved from the crucible by the action of the soda, and will appear in the course of the analysis.

624. The fused mass is treated with water, filtered, and the aqueous solution, which contains chiefly the excess of the carbonate of soda used, and some of the acids of the insoluble mixture must be examined, according to the directions given for the analysis of a mixture soluble in water (585, &c.).

625. The portion of the fused matter which is insoluble in water will generally be found to dissolve, when digested, for a few hours, with the aid of a gentle heat, in DILUTE HYDROCHLORIC or NITRIC ACID; after which the acid solution must be examined, according to the directions given for the analysis of a mixture which is insoluble in water, but soluble in acids (609, &c.).

626. When the insoluble substance has to be examined for alkalies, as in the case of many siliceous minerals, it must be rendered soluble by fusion with carbonate of baryta or lime. As, however, the analysis of such substances is attended with difficulty, the details of the process need not here be considered.[2]

[1] See note to 580.

[2] See Rose, "Analyse Chimique," tom. i, p. 611; ii, 382; also Parnell's "Elements of Chemical Analysis," p. 403.

# PART IV.

## QUANTITATIVE ANALYSIS.

---

*Introductory Remarks.*

627. In the processes which I have now described, the object of the experimenter has been to ascertain what substances are present in a given salt or mixture of salts, which branch of analysis is called *qualitative*. I will now detail a few processes which have for their object the determination of the quantity of the ingredients of saline compounds: this branch of analysis is called *quantitative*. It is not my intention to enumerate the methods which have been devised for the separation and estimation of all, even of the more common compounds, but merely to give the student a general idea of the subject, by conducting him through a few simple examples of quantitative analysis, referring him, if he wishes for more extended information, to the larger works of Rose, Fresenius, and Parnell.[1]

628. I will first briefly describe some of the more important operations which have to be performed in the course of a quantitative analysis; and the student must bear in mind that the more care he bestows upon them, the more correct will be his results; as the loss of a single drop of liquid, or the presence of a very small quantity of soluble matter left in a precipitate, owing to carelessness in washing, will often occasion serious errors.

[1] "Traité Pratique d'Analyse Chimique, par H. Rose," of which an English translation by Dr. Normandy has recently appeared.
"Chemical Analysis, Qualitative and Quantitative," by C. R. Fresenius, translated by Bullock.
"Elements of Chemical Analysis," by E. A. Parnell.

## CHAPTER I.

### OPERATIONS IN ANALYSIS.

*Pulverization.*

629. Most substances may be reduced to sufficiently fine powder for analysis, by pounding in a common Wedgwood mortar; in some cases, however, it is first necessary to break the substance into small fragments, in one of iron or gun metal; or in default of this, the substance may be losely wrapped in strong brown paper, and struck with a hammer. When the substance is difficult of solution, as in the case of some siliceous minerals, it is sometimes necessary to reduce it to an impalpable powder, in a small agate mortar; and on the fineness of this pulverization the success of an analysis often depends.

*Drying.*

630. Many substances, especially when in the state of powder, absorb moisture from the atmosphere, which, of course, adds to their weight. Before weighing out accurately the quantity of the substance for analysis, it is therefore necessary to deprive it of this hygroscopic moisture. This is generally done by heating it in a small basin on the water-bath or sand-bath, care being taken that the heat does not rise so high as to cause decomposition. The hot water box shown in figure 78, is very convenient for drying substances at a low temperature: all the sides are made hollow, and filled with

Fig. 78.

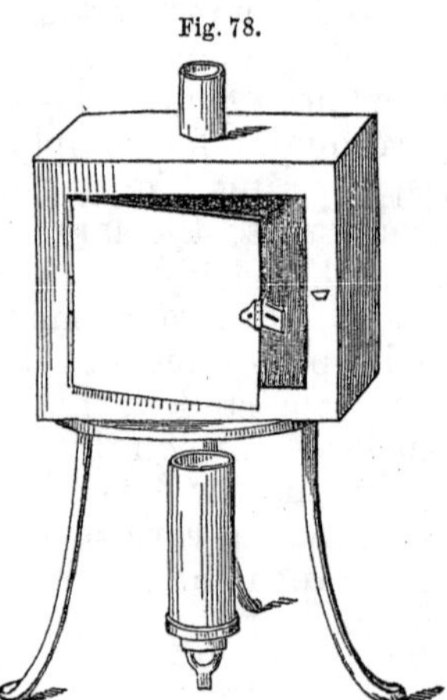

Hot-Water Drying Oven.

water, so that the temperature inside never rises higher than 212°. When a substance thus exposed ceases to lose weight, on being weighed at short intervals, it may be considered sufficiently dry. By using saline solutions, which boil at a higher temperature than water, a steady heat, considerably higher, may be obtained (647).

*Weighing.*

631. Either 20, 25, or 33·3 grains, will generally be found the most convenient quantity to take for quantitative experiments, regard being had to the number of constituents to be estimated, and the quantity of the substance at our disposal. The quantity may depend also on the method we intend to pursue, whether we propose to estimate all the ingredients from the same portion, or from two or more separate portions of the substance. If 20 grains are used, the results, multiplied by five, will give the percentage; or if 25 or 33·3, they must be multiplied by four or three. For most purposes, the student will find a balance that is capable of weighing within one-tenth of a grain, sufficiently accurate; and it should be furnished with weights from one-tenth of a grain to 1000 grains.

A substance should never be weighed while warm, as it causes an upward current of air in its vicinity, which tends to buoy it up, and makes it appear to weigh lighter than it really is. In quantitative analysis, it is, of course, necessary to avoid the slightest loss in the weighed portion, as a deficiency in the weight of the ingredients would be the consequence, and the accuracy of the analysis seriously interfered with. Most substances in the state of fine powder, especially after having been recently ignited, are very prone to absorb moisture from the air; to obviate this, which would add materially to their weight, such substances should be weighed in a covered crucible, as soon as possible after cooling.

When, as is frequently the case, especially with liquids, a substance has to be weighed in a flask, dish, or other vessel, the latter may either be counterpoised with strips of lead or shot, which are conveniently placed in

18

a pill-box; or its weight may be previously noted, and afterwards deducted from the gross weight.

*Solution.*

632. Before the ingredients of a substance can be determined, either qualitatively, or quantitatively, it is necessary to bring the substance into solution. For this purpose, water is to be preferred when the substance dissolves readily in it; and in the case of those compounds which are insoluble in water, one of the acids (generally hydrochloric) is employed, which has been found, in the course of the preliminary examination, to be the best adapted for the purpose. (529–532.)

The solution of a substance is almost invariably assisted by heat, so that it is always advisable to use a vessel for the purpose, which can be heated over a lamp without danger of fracture, as a small Berlin porcelain dish or glass flask. (Figs. 79 & 80.) The latter has the advantage of preventing loss by ebullition or spurting, as any particles of liquid that may be projected from the surface during ebullition, fall against the inner surface, and run back into the flask, especially if it is placed in an inclined position over the lamp. Occa-

Fig. 79.

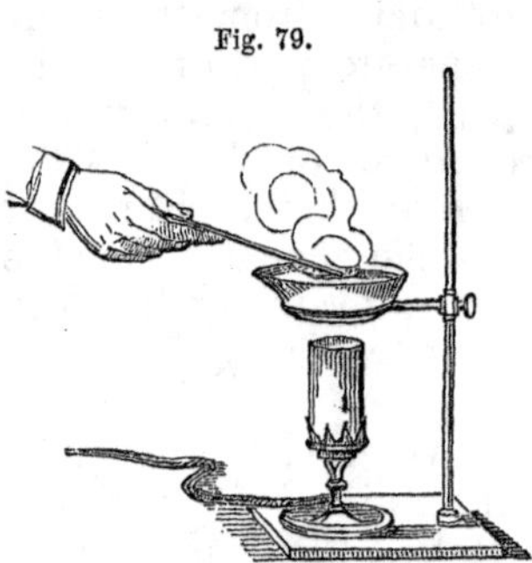

Fig. 80.

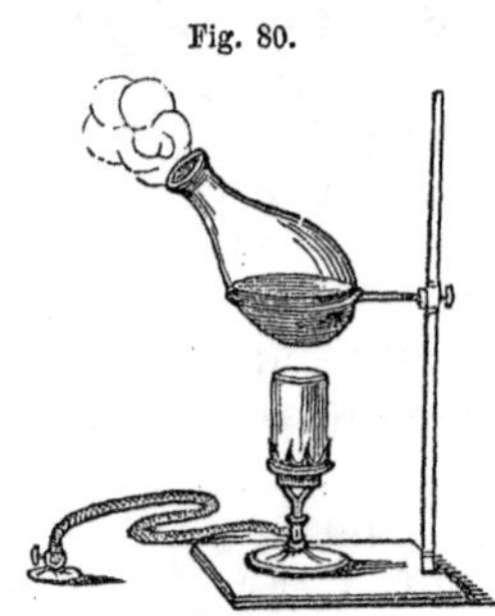

Fig. 81.

sional stirring facilitates the solution, and, as a general rule, the more finely the substance has been pounded, the more readily it dissolves. When a substance has to be digested in acid for a length of time, with the aid of heat, the evaporation of the acid may be in a great measure prevented, by placing a small glass funnel in the mouth of the flask (Fig. 81); the acid condenses, and runs back into the flask.

*Precipitation.*

633. When a substance is obtained in solution, the various compounds present are in most cases separated for the purpose of estimation, by adding to it some solution, which causes one or more of the ingredients to precipitate in the solid state; as, when we wish to estimate the quantity of sulphuric acid in any solution, we add to it a solution of chloride of barium, which, if added in sufficient quantity, causes the whole of the acid to precipitate in the form of sulphate of baryta (403), which, being insoluble in water, may be washed without loss, and when dry is weighed; the weight of the sulphuric acid which it contains, may then be calculated from it (652).

Precipitation is usually effected in upright glasses of the forms shown in Fig. 82. When precipitating a substance in quantitative analysis, it is important that sufficient of the precipitant is added to throw down *the whole* of the substance affected by it, as otherwise a deficiency in weight would be occasioned: this is easily ascertained by adding a drop of the precipitant to the solution filtered from the precipitate, which will cause a further precipitate if sufficient had not before been added. When the precipitate is at all soluble, as the bitartrate of potash, or ammonio-phosphate of magnesia, it is always advisable to allow the mixture to stand several hours before filtering, in order to insure the separation of the whole of the required salt (184).

Fig. 82.

Precipitating Glasses.

When the whole of the precipitate is thrown down, it is separated from the solution either by filtration or decantation (634, 643).

*Filtration.*

634. The process of filtration is that most commonly adopted for separating a precipitate from the solution in which it was formed. The paper best adapted for the purpose is a thin white blotting paper, which should be free from visible holes, and should leave, when burnt, only a minute trace of inorganic matter. Such a paper may be purchased at any of the respectable dealers in chemical apparatus. It is convenient to keep a stock of filters ready cut, of a circular form, and of sizes varying from three to ten inches diameter. These may be made by having circular pieces of tin plate of the different sizes, and scoring round them with a pencil upon the paper, when several sheets may be cut through at once with scissors.

635. The filter, when required for use, is folded twice at right angles (Fig. 83) (66), opened out into a conical

Fig. 83. Fig. 84.

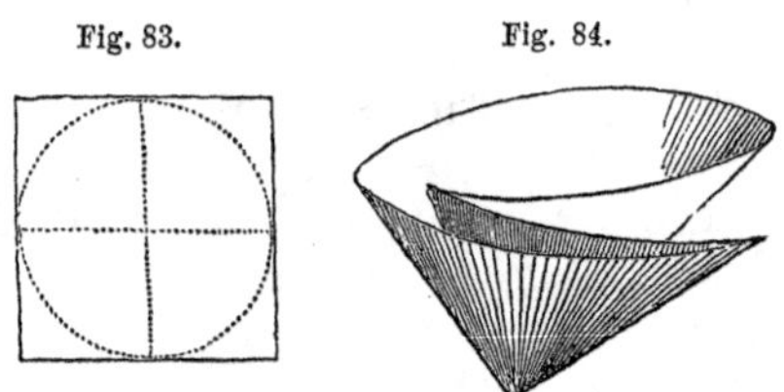

form and placed in a glass funnel, the sloping sides of which should open at an angle of about 60°, when it will be found to match the form of the folded filter, and will support it uniformly throughout. When placed in the funnel, the paper is moistened with water, for the purpose of causing the fibres to expand, and thus diminishing the size of the pores, without, at the same time, choking them with solid particles: if this is not done, and a solution mixed with a precipitate is poured into the dry filter, some of the finely divided particles of the precipitate are drawn into the pores by capillary attraction, and tend to prevent the passage of the clear solution through them. The filter should never be allowed

Fig. 85.

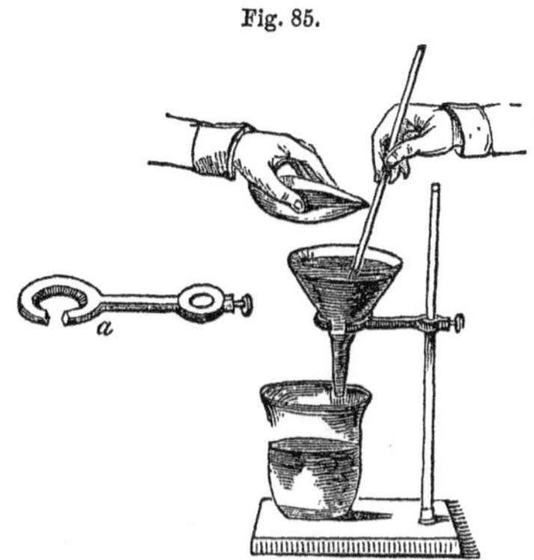


to reach higher than the top of the funnel, as otherwise the weight of the liquid might cause the paper to give way; and there would also be danger of some of the solution running down the outside of the funnel, after passing through the projecting paper. When the filter is thus prepared, it may be supported either on the ring of a retort stand (Fig. 85) (for which the form shown at *a* is very convenient), or on a perforated block of wood placed on the glass intended to catch the filtered solution (169), the hole being made to fit the funnel, as shown in the section (Fig. 86).

Fig. 86.

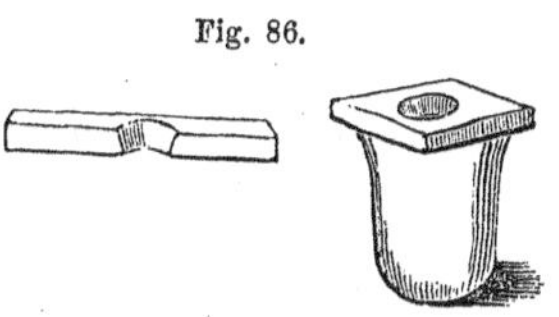

636. The solution to be filtered should be poured gently down a glass rod (Fig. 85), so as to fall on one of the slanting sides of the filter, and not into the apex, as that would endanger the bursting of the paper, and cause splashing. When the whole of the mixture has been poured on the filter, fresh water should not be added for the purpose of washing, until the whole of the solution has passed through; then, by means of a washing-bottle (94), the precipitate left on the filter is well washed; the current of water being applied first towards the upper part of the filter, and directed gra-

dually downwards (Fig. 87). When the filter has been thus nearly filled up with water, allow the whole to run

Fig. 87.

through before adding any more, and then repeat the washing, until a drop of the filtered liquid leaves no fixed residue when evaporated and ignited on platinum foil. If the precipitate, while standing in the filter, cakes together into lumps, these must be broken up by directing upon them a strong current of water from the washing-bottle; as otherwise the water would not penetrate them, and some of the soluble matter would escape removal.

Fig. 88.

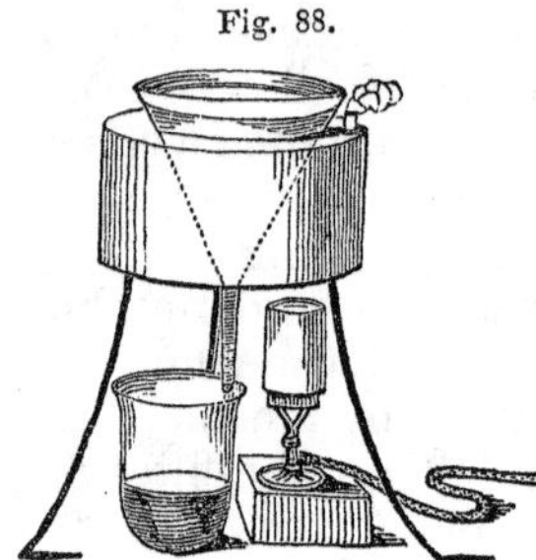

Hot Water Filtering Stand.

637. It is sometimes necessary to keep the mixture hot during filtration, to prevent any of the soluble ingredients solidifying: this may be done very conveniently, by placing the funnel in a zinc or copper box of the form shown in Fig. 88, which may be kept full of hot water, and boiling, if necessary, over a lamp.

638. The liquid is generally filtered into a beaker glass, and

occasionally into flasks or dishes: it is always advisable to cause the stream to run gently down the side of the vessel, and not to fall drop by drop into the centre of the glass, as this would cause splashing and probably some loss. It occasionally happens that some of the precipitate passes through with the filtered solution, as may be seen in the case of freshly precipitated oxalate of lime or sulphate of baryta. When this takes place, it is sometimes necessary to pass it through the filter twice or three times before it comes through quite clear. This may, however, in most cases, be obviated by boiling the mixture before filtering, which causes the finely divided particles of the precipitate to aggregate together. The presence of some saline matters in solution, also, sometimes prevents a precipitate passing through; muriate of ammonia, for example, exerts this property with sulphate of baryta.

639. When the precipitate on the filter is completely washed, the funnel, with its contents, is placed on a small tripod or retort stand, on the warm sand-bath, or near a fire, when the precipitate will gradually dry; it may then be separated from the filter, ignited in a small platinum or porcelain crucible (648), (unless decomposable at a high temperature), and weighed.

640. In cases where the quantity of the precipitate is very small, and where it will not bear a red heat without decomposition, it is often convenient to use two filters for the purpose: they should be folded up together into the proper form, and a hole of about an inch in diameter is then cut with scissors in the centre of the outer one; the inner one is then clipped round the edge, until it weighs exactly the same as the other, when they will, of course, accurately counterpoise each other. They are then again placed one inside the other, the perforated one being outside, after which the mixture may be filtered through them, washed, and dried. When dry, they are separated, and placed in the opposite scales of the balance, when the difference in weight will give the weight of the precipitate, that of the paper being the same in both.

641. The precipitate may also be weighed in a single filter, which should be placed in a covered porcelain

crucible of known weight, dried at 212°, and weighed before the mixture is poured in. When the precipitate on the filter has been thoroughly washed, the latter is placed in the crucible, dried as before, and as soon as it is cold, again weighed; when the increase in weight will, of course, be that of the precipitate.

Fig. 89.

642. It is often necessary, before weighing a precipitate, to burn the filter containing it. After the greater part of the precipitate has been removed, the filter is held with a pair of pliers, and set fire to, over the platinum crucible in which the precipitate is to be ignited, the crucible being placed in a basin, in case any of the ashes should fall over its sides (Fig. 89); these are then collected and ignited in the crucible (648), together with the portion of the percipitate previously removed from the filter, until the whole of the charcoal derived from the paper is burnt away. In cases of great accuracy, the weight of the paper ashes, ascertained by weighing those derived from a similar filter, must be deducted from the gross weight; when the paper is good, however, it does not contain more than one to three thousandths of its weight of inorganic matter, so that this precaution is scarcely necessary in ordinary cases of analysis.

*Decantation.*

643. When a precipitate is found to subside rapidly to the bottom of the liquid, and when it is known to be very insoluble in water, it may be washed by DECANTATION, instead of on a filter, and, in many cases, this is the more expeditious method. The mixture is placed in an upright jar or beaker, which is then filled up with water, and allowed to stand until the precipitate has subsided to the bottom, leaving the superincumbent liquid clear. The latter is then removed with a syphon (Fig. 90), or carefully poured off, and the jar again filled up with distilled water, the process being repeated

until all the soluble matter has been removed. The wet precipitate is then placed upon a filter, or dried in a dish, and weighed.

Fig. 90.

*Evaporation.*

643. The process of evaporation is generally most conveniently effected in Berlin porcelain evaporating basins, either on a sand-bath or over a lanp, a loose cover of filtering-paper being placed over it, if necessary, to prevent particles of dust falling into the liquid. Care must be taken, in quantitative experiments, that no loss is occasioned by spurting, and on this account, it is safer not to allow the liquid absolutely to boil. When a saline solution has to be evaporated to dryness, it often becomes covered, when concentrated, with a pellicle of solid matter, preventing the escape of the steam, which, being thus confined, occasionally causes some of the mixture to be projected violently from the basin. The best way of avoiding this is to stir the mixture constantly with a glass rod, from the time when the pellicle begins to form, until it is evaporated to dryness.

Fig. 91.

644. It is often advisable, and in the case of many liquids, as those containing organic matter, necessary, to evaporate over a water-bath; by this means the heat is never allowed to raise higher than 212°. For this purpose, a common saucepan, or almost any vessel used

for boiling water in, may be employed, placing the dish containing the solution over the top, as shown in Figure 91, so as to expose it to the action of the steam. For the laboratory, a convenient form of water-bath is shown in Figure 92; it may be made of copper or zinc plate,

Fig. 92.

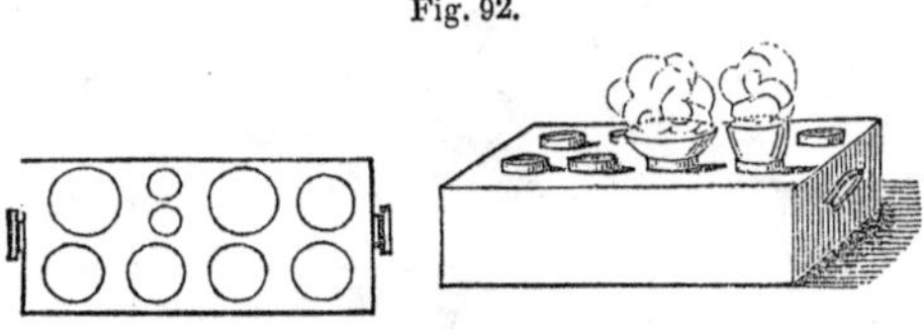

Hot-Water Bath.

and the holes should be fitted with lids, to cover them when not wanted.

646. A convenient method of drying certain substances which are liable to decomposition at a slightly elevated temperature, is to place them under the receiver of the air-pump (Figure 93), over an open pan of strong sulphuric acid; the latter absorbs the moisture which rises from the substance, and a gradual and complete desiccation may be effected at ordinary temperatures. The dishes or tubes containing the substance to be dried, may be placed on a sheet of perforated zinc, resting on the pan of acid. If an air-pump is not at hand, the same effect may be produced, though more slowly, by placing the receiver inclosing the substance and acid, upon a flat piece of glass; or the temporary arrangement shown in Figure 94, may be adopted, *a* being a beaker containing sulphuric acid, over which the substance to be dried is suspended in the dish *e*. Owing to the slowness of the evaporation, this method is

Fig. 93.

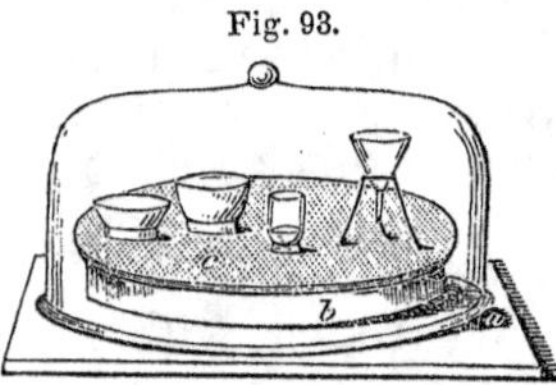


Desiccation over Sulphuric Acid.

Fig. 94.

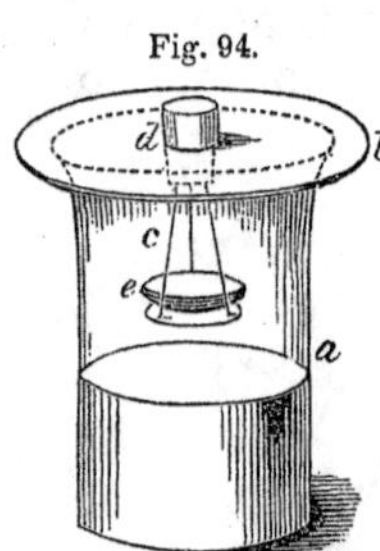


Desiccation over Sulphuric Acid.

well adapted for obtaining large and well-defined crystals from saline solutions, &c.

647. When a uniform temperature is required, higher than 212°, it may be obtained by immersing the dish or flask containing the substance to be evaporated, in a bath of oil or some saline solution, the boiling-point of which is near the desired temperature. Olive oil may be heated to nearly 500° without suffering much decomposition, and forms an extremely useful bath for many purposes, since, by regulating the lamp, and placing a thermometer in the oil, any lower temperature can readily be kept up. The following list shows the boiling-points of a few saturated saline solutions, which will occasionally be found useful for this purpose:—

A saturated solution (at 60°) of

| | | |
|---|---|---|
| Bitartrate of Potash, | boils at | 214° |
| Sulphate of Copper, | " | 216° |
| Chlorate of Potash, | " | 218° |
| Carbonate of Soda, | " | 220° |
| Alum, | " | 220° |
| Borax, | " | 222° |
| Chloride of Sodium, | " | 224° |
| Chloride of Calcium, | " | 230° |
| Tartrate of Potash, | " | 234° |
| Muriate of Ammonia, | " | 236° |
| Nitrate of Potash, | " | 238° |
| Rochelle Salt (Tartrate of Potash and Soda), | " | 240° |
| Nitrate of Soda, | " | 246° |
| Acetate of Soda, | " | 250° |

By adding a further quantity of most of these salts to the *hot* solutions, and thus making them more concentrated, considerably higher temperatures may be obtained.

*Ignition.*

648. It is generally necessary, previous to weighing a precipitate, in quantitative analysis, to heat it to redness, in order to insure perfect dryness. This is usually done in a weighed platinum or porcelain crucible, either in a furnace or over a lamp. When the crucible is to be heated in the furnace or open fire, it should be inclosed in one of earthenware, to protect it from contact with the coals and dirt, a little magnesia

being interposed between the two (note to 580). If the lid is made of the form shown in Figure 95, it may

Fig. 95. Fig. 96.

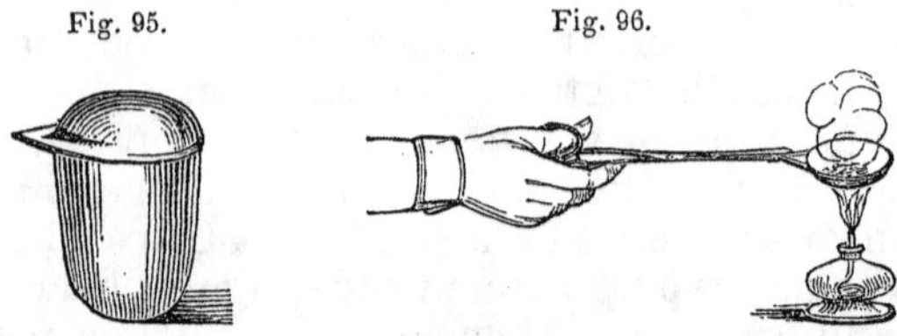

also be used as a capsule, independently of the crucible (Fig. 96).

649. When gas is available, scarcely any other source of heat is necessary for the purpose. A small platinum crucible may be heated to low redness over the naked flame, resting on a small wire triangle placed on the top of the chimney (Fig. 97).

A mixture of gas and air, however, gives a much more intense heat, owing to the more perfect oxidation of the combustible matter: such a mixture is easily obtained by placing a small piece of wire gauze over the chimney, and applying a light to the mixture as it rises through the gauze (Fig. 98). The crucible may be

Fig. 97. Fig. 98. Fig. 99.

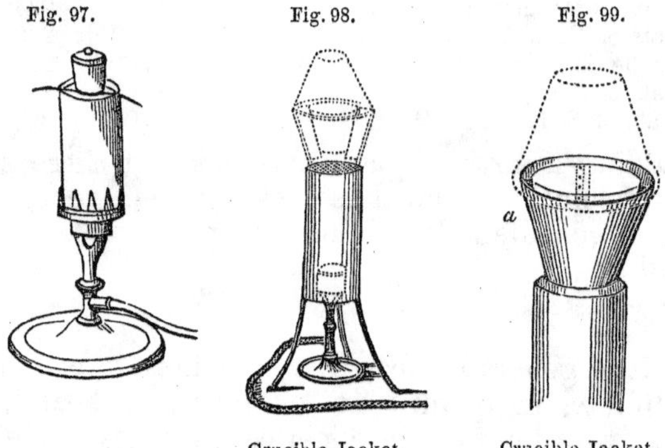


Crucible Jacket. Crucible Jacket.

supported on a wire triangle, or in a *jacket* of thin iron plate, *a* (Fig. 99). In this simple manner, a very powerful lamp furnace is easily obtained, which may be varied in size and form to suit the purposes for which it may be required. Thus, Mr. Hardwich and Mr. Beale have succeeded in making a furnace admirably adapted for heating tubes of a considerable length. An ingenious arrangement is made for gradually increasing the length of the flame along the tube; so that for such purposes as the ultimate analysis of organic substances, it may be substituted, with great advantage, for Liebig's charcoal furnace.

650. Mr. Solly has also contrived a very convenient form of lamp, in which he mixes air with the gas by means of a pair of double bellows, forming a series of blowpipe jets, the combined action of which is very powerful, and is capable of keeping a large platinum crucible almost white hot for any length of time. The gas pipe *a* (Fig. 100) joins the tube *b*, bringing air from the bellows placed underneath, forming in *c* an inflammable mixture; this burns as it issues from the apertures in the burners, the number of which may be multiplied to almost any extent. The crucible may stand on a wire triangle resting on the circle, and the whole may be surrounded by a jacket of thin iron plate to prevent loss of heat by radiation.

Fig. 100.

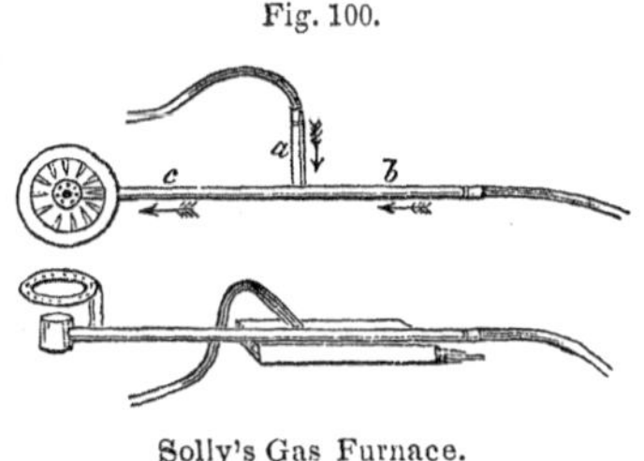


Solly's Gas Furnace.

651. When gas cannot be had, the best lamp for heating a small platinum crucible to redness, is that known as Rose's, the form of which is shown in Figure 101; either alcohol or pyroxylic spirit may be burnt in it.

Fig. 101.

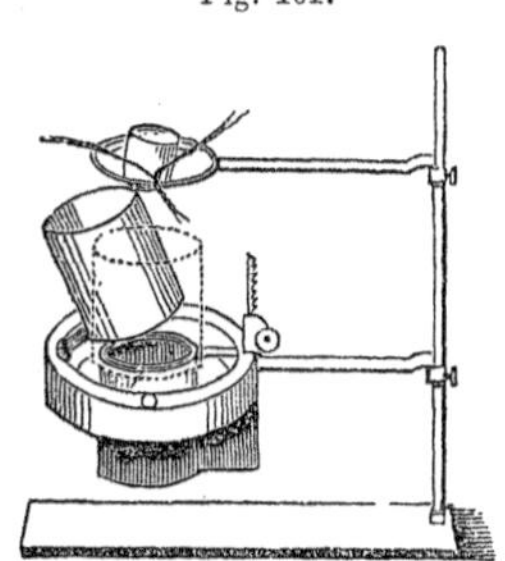

Rose's Spirit Lamp.

*Calculation of Results.*

652. When the weight of a precipitate has been ascertained, it is necessary to calculate that of the constituent whose weight we wish to learn, and this is readily done according to the well-known laws of combination in definite proportion.*

For example, let us suppose that we have to determine the percentage of sulphuric acid ($SO_3$) in dry sulphate of soda ($NaO,SO_3$): we dissolve twenty grains of the salt in water, precipitate the sulphuric acid by means of chloride of barium (403), and weigh the sulphate of baryta thus obtained: from this we have to deduce the weight of the sulphuric acid which it contains; and lastly, to calculate from this the percentage equivalent to it. We find the weight of the sulphate of baryta obtained to be 32·50. Knowing the atomic weight of sulphate of baryta ($BaO,SO_3$) to be 117, and that of sulphuric acid ($SO_3$) to be 40, it is easy to calculate how much of the acid is contained in 32·50 grains of the precipitate, thus:—

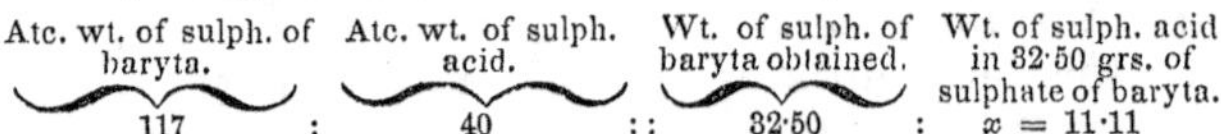


Thus we find that twenty grains of the dry sulphate of soda contain 11·11 of sulphuric acid; and we have now only to reduce it to a percentage, to complete the calculation, thus:—

20 : 11·11 :: 100 : $x$ =55·55 sulphuric acid in 100 parts of dry sulphate of soda. Or as 20 is the fifth part of 100, the same result may be obtained by simply multiplying by 5. 11·11 × 5 =55.55.

---

# CHAPTER II.

## EXAMPLES OF QUANTITATIVE ANALYSIS.

653. In the following examples it is assumed that the nature of the substances has been already ascertained by a qualitative examination; since it is always

* See Fownes' "Manual of Chemistry," p. 187, et seq.

necessary, before proceeding to estimate the constituents of a compound, that we should know what those constituents are.

## SECTION I.

### *Quantitative Analysis of Sulphate of Copper.*
($CuO,SO_3+5Aq.$)

Estimate the quantity of oxide of copper ($CuO$), sulphuric acid ($SO_3$), and water ($HO$), in sulphate of copper.

#### (1.) *Estimation of the Oxide of Copper.*

654. Dissolve twenty grains of the salt in eight or ten ounces of water, in an evaporating basin, and gently boil the solution. Add to it, while boiling, a solution of caustic potash in slight excess, which will throw down the black oxide of copper (370).

$$CuO,SO_3+KO=CuO+KO,SO_3.$$

The mixture is poured upon a filter (636), and carefully washed with boiling water, until the whole of the soluble matter is removed. The precipitate in the filter is then dried, separated from the filter, ignited, and weighed.

When the precipitate consists, as in this case, of the substance whose weight we wish to ascertain, uncombined with other matter, we have only to reduce the amount thus obtained, to a percentage:—

20 : weight of the precipitate of oxide : : 100 : $x$ = : percentage of oxide of copper. Or, in other words, multiply the weight by 5.

#### (2.) *Estimation of the Sulphuric Acid.*

655. The acid may be estimated either from the same portion of the substance as was used in determining the oxide (in which case the solution filtered from the precipitated oxide must be concentrated by evaporation (644)), or a fresh portion of the salt may be used. The latter method is in this case the simplest.

656. Dissolve twenty grains of the sulphate in water as before, acidify it with a few drops of nitric acid, and

add a solution of chloride of barium (*BaCl*) as long as it causes any precipitate. Sulphate of baryta ($BaO,SO_3$) is thus thrown down (403), and the whole of the sulphuric acid is in this way removed from the solution. As the mixture in its present state would not filter well (638), it is advisable to boil it before filtering, when it will be found that the solution will pass through clear. The precipitate is well washed with hot water on the filter, dried, ignited in a platinum or porcelain crucible, and weighed. Then, knowing the atomic weight of sulphate of baryta to be 117, and that of sulphuric acid ($SO_3$) to be 40, or, in other words, that every 117 parts of the former contain 40 of the latter, it is easy to calculate the quantity of sulphuric acid contained in the precipitate, thus:—

$$117 : 40 :: \text{weight of the precipitate} : x = \left\{ \begin{array}{c} \text{The sulphuric acid in 20 grains of} \\ \text{sulphate of copper.} \end{array} \right\}$$

which number, multiplied by five, will represent the percentage of sulphuric acid in crystallized sulphate of copper.

(3.) *Estimation of the Water.*

657. The water is estimated by heating 20 grains on the sand-bath in a counterpoised crucible, at a temperature of about 400°, until it ceases to lose weight: in this way the water is expelled, and its quantity is shown by the loss of weight, which, when multiplied by five, will give the percentage of water.*

## SECTION II.

### *Quantitative Analysis of Chloride of Potassium* (KCl).

Estimate the quantity of potassium and chlorine in the chloride.

(1.) *Estimation of the Potassium.*

658. Dissolve twenty grains of the salt in as small a

* When the quantity of water has to be estimated in salts which cannot bear the necessary heat without the volatilization of a portion of their acid, the salt should be intimately mixed with five or six times its weight of protoxide of lead, or some other strong base, before exposure to heat: this combines with, and fixes, any of the acid that may be disengaged from the other base.

quantity of water as possible, in an evaporating basin; add bichloride of platinum to the solution, and evaporate the mixture to dryness on a water-bath (645). Treat the residue with alcohol, to dissolve out the excess of the bichloride, and wash the insoluble double chloride of platinum and potassium (185) with fresh alcohol, on a weighed filter (641). If it is found that the alcohol which is added to the evaporated residue, does not become sensibly colored yellow, it is owing to there being no excess of chloride of platinum present; that salt having been added in too small quantity. In this case, a little more chloride of platinum should be added, and the mixture again evaporated as before, and subsequently treated with alcohol. The precipitate is then dried on the filter, at a moderate heat, and weighed.

659. The atomic weight of the double chloride (KCl, $PtCl_2$) being 247, and that of potassium being 40, we deduce the quantity of potassium contained in the twenty grains of the chloride, thus:—

247 : 40 : : {Wt. of double chloride of platinum and potassium} : {Wt. of potassium contained in 20 grains of the chloride.}

and lastly, it is reduced to a percentage, thus:—

20 : {Wt. of potassium contained in 20 grains of the chloride} : : 100 : percentage of potassium.

or the same result may be obtained by multiplying by five.

### (2.) *Estimation of the Chlorine.*

660. Twenty grains of the chloride are dissolved in three or four ounces of water, as before; the solution is then heated, acidified with a few drops of nitric acid, and treated with a solution of nitrate of silver, as long as it causes any precipitate (429, 633).

$$KCl + AgO,NO_5 = \mathrm{AgCl} + KO,NO_5.$$

The mixture is then boiled for a few minutes (as otherwise a portion of the precipitate would pass through the pores of the paper (638) ) and filtered. The precipitated chloride of silver is thoroughly washed with distilled water on the filter, and dried; it is then removed from the paper, and gently ignited in a counterpoised porcelain crucible, until it fuses into a waxy mass, and weighed.

661. The atomic weight of chloride of silver is 144, and that of chlorine 36, so that we deduce the weight of the chlorine from that of the chloride, thus:—

144 : 36 : : wt. of the chloride of silver : { Wt. of chlorine contained in 20 grains of chloride of potassium. }

Then, for the percentage, multiply by five.

## SECTION III.

*Quantitative Analysis of a mixture of Sulphate of Copper* ($CuO,SO_3+5Aq$) *and Chloride of Sodium* ($NaCl$).

Estimate the quantity of oxide of copper ($CuO$), sodium, sulphuric acid ($SO_3$), chlorine, and water in the mixture.

(1.) *Estimation of the Water.*

662. The water is estimated in the manner described in paragraph (657).

(2.) *Estimation of the Oxide of Copper.*

663. Dissolve twenty-five grains of the mixture in water, acidify the solution with a few drops of hydrochloric acid, put it into a beaker, and pass through it a stream of hydrosulphuric acid, until it is saturated (701); the whole of the copper is in this way thrown down as sulphide (368). Filter, and wash the precipitate with distilled water, which should contain in solution a little hydrosulphuric acid, as otherwise a trace of copper is liable to become oxidized and dissolved. The clear solution, together with all but the last washings, is set aside for subsequent examination (664). The washed precipitate of sulphide of copper is now for the most part separated from the filter, which latter is to be burnt, and the ashes added to the rest of the precipitated sulphide. The precipitate, containing the whole of the copper, is then digested in strong nitric acid, until the whole of the precipitate is dissolved, or until nothing remains undissolved but a little sulphur, of a pale yellow color. The acid solution thus obtained is diluted with water, and filtered, if necessary, from the

undissolved sulphur. The whole of the copper is now contained in the solution as nitrate. The copper is next to be thrown down as oxide by potash, dried and weighed, in the manner already described (654). The percentage is calculated as follows:—

25 : wt. of oxide obtained : : 100 : percentage of oxide of copper in the mixture. Or, as 25 is the fourth part of 100, the same result is arrived at by simply multiplying the weight of the oxide obtained, by 4.

(2.) *Estimation of the Sodium.*

664. The chloride of sodium contained in the solution filtered from the sulphide of copper, is concentrated by evaporation, and then converted into sulphate of soda ($NaO,SO_3$), by evaporating to dryness with a slight excess of strong sulphuric acid.

$$NaCl + HO,SO_3 = \mathrm{NaO,SO_3} + HCl.$$

The residue is gently ignited in a counterpoised covered crucible, in which a fragment of carbonate of ammonia is suspended by means of a strip of platinum foil, and weighed; the weight of the sodium is thus calculated:—

| At. wt. of sulph. soda. | | At. wt. of sodium. | | Wt. of sulph. of soda obtained. | | Sodium in 25 grs. of mixture. |
|---|---|---|---|---|---|---|
| 72 | : | 24 | : : | $a$ | : | $x$ |

which when multiplied by four, gives the percentage of sodium.

(3.) *Estimation of the Sulphuric Acid.*

665. A second portion of twenty-five grains of the mixture is dissolved in water, for the purpose of estimating the sulphuric acid and chlorine.

Add to it first, a solution of nitrate of baryta, as long as it causes any precipitate, and boil the mixture for a few minutes to prevent any of the finely divided sulphate of baryta passing through the filter (638). The precipitate is washed, dried, and weighed, the clear solution being reserved for estimating the chlorine (666); the quantity of sulphuric acid is then calculated in the manner already described (656), twenty-five being substituted for twenty in the calculation.

(4.) *Estimation of the Chlorine.*

666. The solution filtered from the sulphate of baryta (665), together with the first washings, is heated in an evaporating basin, over a lamp, and treated with nitrate of silver as long as it causes any precipitate (429). The chloride of silver thus formed is filtered, dried, and weighed ; and the weight of the chlorine deduced from it in the manner described in (660), twenty-five being substituted for twenty in the calculation.

Thus the percentage of oxide of copper, sodium, sulphuric acid, chlorine, and water, will have been ascertained.

## SECTION IV.

*Quantitative Analysis of a mixture of Sulphate of Zinc* ($ZnO, SO_3+7Aq$) *and Carbonate of Baryta* ($BaO,CO_2$).

Estimate the quantity of water, sulphuric acid, oxide of zinc, carbonic acid, and baryta, in the mixture.

667. Twenty grains of the mixture are to be boiled with three or four ounces of water, which will dissolve out the sulphate of zinc from the carbonate of baryta. The mixture is filtered, and the insoluble portion washed until the washings leave no residue when evaporated on platinum foil. The solution contains the whole of the sulphate of zinc, while the carbonate of baryta remains undissolved; the latter is retained for subsequent examination (671).

(1.) *Estimation of the Oxide of Zinc.*

668. The solution is treated with carbonate of potash as long as it causes any precipitate, and then boiled. The zinc is thus thrown down as a basic carbonate (257); the precipitate is washed on a filter, dried, and ignited, when the water and carbonic acid of the compound are expelled, and pure oxide of zinc ($ZnO$) remains, which is weighed; the weight thus obtained, multiplied by five, gives the percentage of oxide of zinc.

(2.) *Estimation of the Sulphuric Acid.*

669. The solution filtered from the precipitated carbonate of zinc, is now supersaturated with nitric acid, and boiled to expel the carbonic acid; chloride of barium is added, and the mixture is boiled for a few minutes to clarify it (638); the sulphate of baryta thus obtained is washed, dried, ignited, and weighed as already described (656), and the weight of the sulphuric acid which it contains, calculated as before.

(3.) *Estimation of the Water.*

670. The water may be determined by heating the mixture on the sand-bath, as described in (657).

(4.) *Estimation of the Baryta.*

671. The residue which was insoluble in water (667), is now removed from the filter into a small beaker, and dissolved in dilute hydrochloric acid, a gentle heat being applied, if necessary. In this way, the carbonate is decomposed, carbonic acid is given off, and chloride of barium remains in solution.

$$BaO,CO_2 + HCl = BaCl + HO + CO_2.$$

The solution thus obtained is treated with dilute sulphuric acid, as long as any precipitate is produced, and then boiled: the precipitate is separated by filtration, washed, dried, ignited, and weighed.

The atomic weight of sulphate of baryta being 117, and that of baryta being 77, the quantity of the latter in the precipitate is calculated as follows:—

117 : 77 : : wt. of sulphate of baryta obtained : {Wt. of baryta contained in 20 grains of the mixture.}

(5.) *Examination of the Carbonic Acid.*

672. The carbonic acid is estimated by decomposing the carbonate with hydrochloric acid in a flask, and determining the amount of loss, as described in paragraph (174).

## SECTION V.

*Quantitative Analysis of Magnesian Limestone; consisting of Carbonate of Lime* ($CaO,CO_2$), *Carbonate of Magnesia* ($MgO, CO_2$), *Peroxide of Iron* ($Fe_2O_3$), *a little Silica* ($SiO_3$), *and moisture.*

---

Determine the quantity of lime, magnesia, peroxide of iron, carbonic acid, silica, and moisture, in magnesian limestone. Reduce about 100 grains of the mineral to moderately fine powder.

---

### (1.) *Estimation of the Hygroscopic Moisture.*

673. Weigh fifty grains of the pounded mineral in a small counterpoised crucible or evaporating dish, and dry it on a water-bath, or on the hot part of the sand-bath: weigh it at intervals of a quarter or half an hour, until it ceases to lose weight (630). The loss will be the quantity of moisture in fifty grains, which, when multiplied by two, will give the percentage.

### (2.) *Estimation of the Silica.*

674. Weigh twenty-five grains of the pounded mineral in a counterpoised flask, moisten it with a little water, and add dilute hydrochloric acid in small quantities, to avoid too violent effervescence (419). When the greater part is dissolved, warm it with a little fresh acid, which will dissolve everything but the small quantity of silica. The mixture is now filtered, and the precipitate thoroughly washed, the solution being retained for subsequent examination (675); the filter containing the precipitate is then ignited and weighed. The weight of the siliceous residue, multiplied by four ($25 \times 4 = 100$), gives the percentage of silica in the stone.

### (3.) *Estimation of the Peroxide of Iron.*

675. The acid solution filtered from the siliceous residue, is now neutralized with ammonia, and a few drops of hydrosulphate of ammonia are added, which will throw down the iron as the black sulphide (279).

This is to be filtered and carefully washed, the solution being retained for further examination (676); when the whole of the soluble matter is removed, the filter, with the moist precipitate, is digested in hydrochloric acid, until the black sulphide is entirely decomposed, and nothing but sulphur remains undissolved. The solution is now diluted, and separated by filtration from the sulphur and fragments of the first filter, which must be well washed to remove the whole of the soluble matter. The solution is well boiled to expel the hydrosulphuric acid, and then heated with a little nitric acid, for the purpose of peroxidizing the iron. Ammonia is now added in slight excess, which precipitates the whole of the iron as hydrated peroxide (280). The precipitate is filtered, dried, ignited, and weighed. The weight, multiplied by four, gives the percentage of peroxide of iron in the mineral.

(4.) *Estimation of the Lime.*

676. The solution filtered from the sulphide of iron, is now to be boiled with a slight excess of hydrochloric acid, to decompose the excess of hydrosulphate of ammonia, and expel the hydrosulphuric acid: when the smell of that gas is no longer perceptible, filter the solution, if necessary, from any sulphur that may have been precipitated, and neutralize the clear solution with ammonia; after which add oxalate of ammonia as long as it causes any precipitate (633). This throws down the lime as oxalate (218), while the magnesia remains in solution, not being precipitated by oxalate of ammonia in the presence of muriate of ammonia, which is contained in the solution. The mixture is boiled (638), and filtered, and the precipitate washed and dried; the solution being retained for subsequent examination (677). The oxalate of lime is now removed from the filter, and ignited, by which means it is decomposed, and converted into carbonate of lime: at a red heat, however, a portion of the newly formed carbonate is decomposed, the carbonic acid being expelled, and thus leaving a little caustic lime (CaO) mixed with the carbonate. When cool, it is moistened with a

solution of carbonate of ammonia, and again gently heated, to expel the excess of carbonate of ammonia: the whole of the lime is thus converted into carbonate, in which state it is weighed.

The atomic weight of carbonate of lime being 50, and that of lime 28, the weight of the latter is thus calculated:—

50 : 28 : : wt. of carbonate obtained : {Wt. of lime contained in 25 grains of the mineral.}

which number, multiplied by four, gives the percentage of lime.

(5.) *Estimation of the Magnesia.*

677. The solution filtered from the oxalate of lime is now to be concentrated by evaporation, and treated with a mixture of phosphate of soda and a considerable excess of caustic ammonia; when the mixture is set aside for a few hours, being stirred at intervals with a glass rod. The magnesia is thus thrown down as the double phosphate of ammonia and magnesia (206). Before filtering, a little of the mixture should be tested with a few drops more phosphate of soda, and allowed to stand a short time; when, if no fresh precipitate is formed, it may be concluded that a sufficient quantity of the precipitant has been added.

The mixture is now filtered, washing always with water containing a little free ammonia, as the double phosphate is slightly soluble in pure water (any of the precipitate that adheres to the sides of the glass being separated by means of a feather, which must be well washed from all adhering particles), and the precipitate, after drying, is ignited in a small counterpoised platinum or porcelain crucible, and weighed. During ignition, the double phosphate is decomposed into phosphate of magnesia ($2MgO,PO_5$), the water and ammonia being driven off (206).

The atomic weight of the phosphate of magnesia thus formed, being 112, and that of magnesia 20, the quantity of magnesia contained in the precipitate is thus calculated;—

112 : 40[1] : : wt. of phosphate : wt. of magnesia in 25 grs. of the mineral,

which when multiplied by four, gives the percentage of magnesia in the mineral.

(6.) *Estimation of the Carbonic Acid.*

678. The carbonic acid is estimated in the manner described in paragraph (174).

679. Thus we shall have determined the percentage of the—

| | |
|---|---|
| Water, . . . . . . | |
| Silica, . . . . . . | |
| Peroxide of iron, . . . . . | |
| Lime, . . . . . . | |
| Magnesia, . . . . . . | |
| Carbonic acid, . . . . . | |
| Loss, . . . . . . | |
| | 100·00 |

which when added together, ought to amount to about 99 or 99·5 ; and the small deficiency, always inevitable in such analyses, is put down as "loss," to make up the 100 parts.

## SECTION VI.

*Quantitative Analysis of Copper Pyrites ; consisting of Copper, Iron, Sulphur, Silica (Quartz), and moisture.*

Determine the quantity of copper, iron, sulphur, silica, moisture in copper pyrites.

(1.) *Estimation of the Moisture.*

680. Dry 100 grains of the pounded pyrites on the sand-bath, in a counterpoised crucible or dish : the loss is the percentage of moisture.

681. Boil 33·33 grains of the pounded mineral in aqua regia (698) until the sulphur which remains undissolved, collects into a yellowish porous lump. Dilute the acid mixture with two or three times its bulk of water; filter, and wash the insoluble residue (consisting of sulphur

[1] 40=20×2 ; because each equivalent of the phosphate ($2MgO,PO_5$) contains two equivalents of magnesia.

and silica) until the whole of the soluble matter is separated from it (636); reserve the insoluble residue for further examination (684).

682. As the nitric acid present in the aqua regia, would interfere with the action of the hydrosulphuric acid made use of in a subsequent stage of the analysis, it should be expelled by evaporation with a slight excess of hydrochloric acid (562).

### (2.) *Estimation of the Sulphur.*

683. A portion of the sulphur will have become oxidized by the action of the nitric acid in the aqua regia; and the sulphuric acid thus formed, will be contained in the filtered liquid, in combination with the oxides of copper and iron.

Add chloride of barium to the solution as long as it causes a precipitate (403); boil the mixture; filter, wash, and ignite the precipitate. From the weight of the sulphate of baryta thus obtained, that of the sulphur from which the sulphuric acid was derived, may be ascertained by the following calculation:—

$$\underbrace{117}_{\text{Atc. wt. of sulphate of baryta.}} : \underbrace{16}_{\text{Atc. wt. of sulphur.}} :: \underbrace{a}_{\text{Wt. of sulphate of baryta obtained.}} : \underbrace{x}_{\text{Wt. of sulphur dissolved.}}$$

684. The weight of the sulphur which resisted the action of the aqua regia (681), must now be estimated. The undissolved residue is to be thoroughly dried at 212°, and weighed: it is then gradually heated to redness in a counterpoised or weighed porcelain crucible, until the whole of the sulphur is burnt off, when it must be again weighed, the loss during ignition being of course the sulphur. This weight must be added to that already deduced from the sulphate of baryta (683), which, together, will give the quantity of sulphur contained in 33·33 grains of the mineral, or one-third of the percentage.

### (3.) *Estimation of the Silica.*

685. The siliceous matter is left after the expulsion of the sulphur, by ignition, from the residue insoluble

in aqua regia.[1] The percentage is obtained by multiplying by three.

(4.) *Estimation of the Copper.*

686. The solution filtered from the sulphate of baryta (683), containing a slight excess of chloride of barium, is now treated with a slight excess of sulphuric acid, to remove the superfluous baryta, which is separated by filtration. The clear liquid is then subjected to a current of hydrosulphuric acid gas as long as any precipitate is produced, and filtered. The sulphide of copper thus formed is treated in the manner described in (663), and from the weight of the oxide (CuO), that of the copper is calculated thus:—

| Atc. wt. of oxide of copper. | | Atc. wt. of copper. | | Wt. of oxide obtained. | | Wt. of copper in 33·33 grs. of pyrites. |
|---|---|---|---|---|---|---|
| 40 | : | 32 | : : | $a$ | : | $x$ |

which, when multiplied by three, represents the percentage of copper in the mineral.

(5.) *Estimation of the Iron.*

687. The solution filtered from the sulphide of copper, must now be boiled to expel the excess of hydrosulphuric acid, filtered, if necessary, from sulphur, and afterwards heated with a little nitric acid, for the purpose of peroxidizing the iron (269). Ammonia is added in slight excess: this throws down the iron as hydrated peroxide (280), which is to be filtered, dried, ignited, and weighed. The weight of the iron contained in the precipitate is thus calculated:—

| Atc. wt. of peroxide of iron. | | Atc. wt. of iron. | | Wt. of oxide obtained. | | Wt. of iron contained in 33·33 grs. of pyrites. |
|---|---|---|---|---|---|---|
| 40 | : | 28 | : : | $a$ | : | $x$ |

[1] A trace of tin is occasionally found in the pyrites, and will be contained in this residue as peroxide ($SnO_2$). It may be detected by the blowpipe (379), and if found to be present, the residue is boiled with hydrochloric acid to dissolve out the minute globules of metallic tin; the chloride thus formed (SnCl) is filtered, and converted into peroxide by boiling with nitric acid; the excess of acid is then expelled by evaporation, when the peroxide of tin, if present in sufficient quantity, may be weighed, and its weight deducted from that of the siliceous residue.

688. The quantities thus obtained, should, when added together, amount to a fraction less than 100, the deficiency being, as before, set down as "loss":—

| | |
|---|---|
| Copper, . . . . . . . | |
| Iron, . . . . . . . . | |
| Sulphur, . . . . . . . | |
| Silica, . . . . . . . . | |
| Tin (?), . . . . . . . | |
| Moisture, . . . . . . . | |
| Loss, . . . . . . . | |
| | 100·00 |

# PART V.

---

## CHAPTER I.

### REAGENTS.

689. The following is a list of the reagents, &c., usually employed in testing and analysis :—

- Sulphuric acid, strong and dilute.
- Hydrochloric acid.
- Nitric acid.
- Nitrohydrochloric acid (aqua regia).
- Oxalic acid.
- Acetic acid.
- Tartaric acid.
- Hydrosulphuric acid (sulphuretted hydrogen).
- Ammonia.
- Hydrosulphate of ammonia.
- Carbonate of ammonia.
- Oxalate of ammonia.
- Phosphate of soda and ammonia (microcosmic salt).
- Potash.
- Carbonate of potash.
- Nitrate of potash.
- Iodide of potassium.
- Chromate of potash.
- Cyanide of potassium.
- Ferrocyanide of potassium (yellow prussiate of potash).
- Ferridcyanide of potassium (red prussiate of potash).
- Antimoniate of potash.
- Carbonate of soda.
- Phosphate of soda.
- Borax.
- Lime water.
- Sulphate of lime.
- Chloride of calcium.
- Chloride of barium.
- Nitrate of baryta.
- Perchloride of iron.
- Nitrate of cobalt.
- Sulphate of copper.
- Ammonio-sulphate of copper.
- Acetate of lead.
- Subacetate of lead.
- Nitrate of silver.
- Ammonio-nitrate of silver.
- Perchloride of mercury.
- Protochloride of tin.
- Perchloride of gold.
- Bichloride of platinum.
- Sulphate of indigo.
- Solution of starch.
- Black flux.
- Distilled water.
- Alcohol.
- Litmus and turmeric paper.

690. Most of these substances, as they are met with in commerce, being always more or less impure; and,

as those even which are sold in the shops as pure reagents, are not unfrequently found, on examination, to be otherwise; it is always necessary, before taking a reagent into use, to ascertain by experiment whether it is of sufficient purity for the purposes for which it is intended. It may be stated as a general rule, that, when a chemical substance is required for use in analysis, it ought to be as nearly pure as possible; while, for the other operations of chemistry, the substances which are usually met with in commerce are sufficiently pure. The following brief remarks relative to the more common impurities of reagents, together with their principal uses, will probably be found useful to the student.

*Sulphuric Acid* ($HO,SO_3$).

691. Sulphuric acid, as found in commerce, is never pure. The most common impurities are sulphate of lead ($PbO,SO_3$), nitric acid ($NO_5$), or binoxide of nitrogen ($NO_2$), and occasionally arsenic, and other saline matters.

(*a*) If it contains the first, it will become turbid when diluted with four or five times its bulk of water, owing to the sulphate of lead, which is soluble in the strong acid, being insoluble in the dilute.

(*b*) Nitric acid, or the binoxide of nitrogen, is detected by warming a little of the acid in a test-tube, with a small crystal of protosulphate of iron (449); or by boiling a small portion tinged with a solution of sulphate of indigo, when, if nitric acid is present, the blue color will disappear (452).

(*c*) Arsenic may be detected by Marsh's test (313).

(*d*) Any fixed saline impurity remains as a residue when a few drops of the acid are evaporated on platinum foil.

692. The uses of sulphuric acid are very numerous. Besides being employed extensively in many branches of manufacture, it is used in the laboratory as a powerful decomposing agent; owing to its strong affinity for bases, nearly all saline compounds are decomposed by it, and its solvent powers are also very great. It is often employed for the purpose of decomposing organic mat-

ter; also in the preparation of hydrogen, hydrosulphuric acid, and other gases; as a test for certain metals, and for many other purposes.

693. When *dilute* sulphuric acid is required, it is prepared by mixing together, in a porcelain basin, one part of the strong acid with four parts of distilled water, always *adding the acid to the water*, which should be kept constantly stirred, and allowing the precipitated sulphate of lead (if any) to subside, after which the clear liquid may be poured off.

*Hydrochloric Acid* ($HCl$).

694. This acid, in the form met with in commerce, is never pure, usually containing sulphuric acid and chloride of iron, and occasionally free chlorine and traces of arsenic.

(*a*) Evaporate a drop or two on platinum foil: if pure, no residue is left.

(*b*) Dilute a portion with four or five times its bulk of distilled water, and add a drop of chloride of barium: if sulphuric acid is present, a white precipitate is produced (403, 428).

(*c*) Add ammonia in excess: a brown precipitate indicates iron (280).

(*d*) Boil a little of the acid, tinged with sulphate of indigo: if it contains free chlorine, the blue color is bleached.

(*e*) Arsenic may be detected by Marsh's test (313).

695. The uses of hydrochloric acid are very numerous, especially in analysis, in which it is of constant value as a solvent for substances which are insoluble in water; most of the metals dissolve readily in it, forming soluble chlorides, and it is occasionally used to precipitate silver and mercury from their solutions.

When *dilute* hydrochloric acid is required, the strong acid may be diluted with about twice its bulk of water.

*Nitric Acid* ($HO,NO_5$).

696. Nitric acid, as met with in commerce, usually contains sulphuric and hydrochloric acids, and occasionally a little fixed saline matter.

(*a*) The latter may be detected by evaporating a few drops on platinum foil, when any fixed impurities will be left.

Dilute a little of the acid with water, and divide it into two portions.

(*b*) To the first, add chloride of barium; if a white precipitate is produced, sulphuric acid is present (403, 445).

(*c*) To the other, add nitrate of silver: a white precipitate, soluble in ammonia, indicates hydrochloric acid (429).

697. Nitric acid is used chiefly as a solvent for substances which are insoluble in water, especially some of the metals, which it readily oxidizes, and converts into nitrates, nearly all of which are soluble in water. It is, also, frequently employed to raise compounds to a higher state of oxidation, as in converting the protoxide of iron ($FeO$) into the peroxide ($Fe_2O_3$).

When *dilute* nitric acid is required, it may be prepared by mixing one part of the strong acid with two parts of distilled water.

### *Nitrohydrochloric Acid (Aqua Regia).*

698. This is always prepared when required, by mixing together strong nitric and hydrochloric acids, usually in the proportion of one part of nitric to four of hydrochloric. Its chief uses depend on its intense oxidizing or chlorinizing properties, whereby the most refractory metals, some of which resist the action of all other acids, are brought into solution.

### *Hydrosulphuric Acid* (HS). (*Sulphuretted Hydrogen.*)

699. This reagent, whether required in the gaseous form or in solution, is always prepared in the laboratory. Fragments of sulphide (sulphuret) of iron (FeS) are placed in a bottle, *a* (Fig. 102), and treated with dilute sulphuric acid (which for this purpose should consist of one part of acid and eight parts of water), which disengages the gas.

$$FeS + HO,SO_3 = FeO,SO_3 + HS.$$

The gas thus formed, is passed through water contained in the second bottle, *b*, for the purpose of purifying it

Fig. 102.

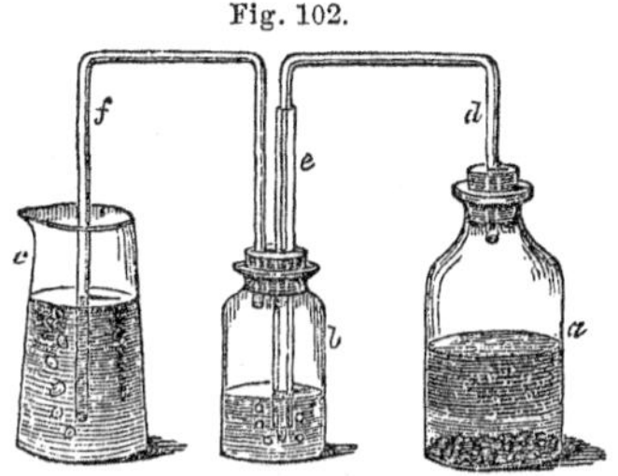


Sulphuretted Hydrogen Apparatus.

from any sulphuric acid and iron that may have been carried over mechanically, and is then conducted, by the bent tube, *f*, into a bottle of distilled water, when an aqueous solution of the gas is required, or into a jar containing any solution which it is intended to act upon (95).

It must be borne in mind, when experimenting with this gas, that it is not only highly offensive, but poisonous, and induces headache and nausea even when largely diluted with air: on this account it should be prepared either near a ventilating flue, or in the open air; never in a close room.

700. In most cases of mere testing, the aqueous solution is the most convenient form in which to apply it. The water should be saturated with the gas, of which it is capable of retaining in solution about its own volume; this may be judged of by its strong disagreeable smell, resembling that of rotten eggs, and by its giving a copious white precipitate of sulphur when treated with perchloride of iron (278). It should also be tested for iron, which it sometimes contains when carelessly prepared; if such is the case, it becomes dark colored on the addition of ammonia, owing to the formation of sulphide of iron (271). The solution should not be kept long, as it is liable to decompose, unless carefully closed from the air, the oxygen of which combines with the hydrogen to form water (*HO*), while sulphur is deposited.

701. When it is required to precipitate, by hydrosulphuric acid, *the whole* of any metal in a solution, it is necessary to pass the gas at once into it; and this should be continued until the liquid is saturated, which is known by removing the gas-delivering tube, and blowing away the superincumbent air; when, if it smells distinctly of the gas, the solution may be considered saturated, and the whole of the metal must have been converted into sulphide.

702. The important uses to which hydrosulphuric acid is applied, render it of great value in many processes of analysis. It precipitates many of the metals from their solutions as insoluble sulphides; and is one of the reagents employed in determining the class to which an unknown metal in solution belongs (179). It is also extensively used in quantitative analysis, on account of the perfect manner in which it separates the whole of many of the metals from their solutions. Hydrosulphuric acid is also sometimes useful as a deoxidizing agent, reducing metallic oxides in solution to a lower degree of oxidation, as the peroxide of iron to the protoxide; this property is owing to the facility with which it is decomposed, yielding up its hydrogen to the oxygen of the oxide, while the sulphur is usually set free (278).

*Oxalic Acid* ($HO,C_2O_3$).

703. Oxalic acid occasionally contains traces of nitric acid (which causes it to deliquesce in damp air, and to have a slightly acid smell), and also fixed saline matter.

(*a*) The first may be detected by boiling the solution with a drop or two of sulphate of indigo (452).

(*b*) The latter, if present, is left as a fixed residue after ignition on platinum foil.

It may be easily purified by recrystallization.

704. The chief use to which oxalic acid is applied in analysis, is to precipitate lime from its solutions (218). (See also *Oxalate of Ammonia* (714).) For use as a test, one part of the crystallized acid may be dissolved in ten parts of water; but, as the solution is liable to decompose, it is better to keep it in the solid state, and to dissolve a little when wanted.

*Acetic Acid* $(HO, C_4H_3O_3)$.

705. This acid is often contaminated with one or more of the following substances: sulphuric, sulphurous, hydrochloric, and nitric acids, lead, and other saline matter.

(*a*) Any fixed impurity may be detected by heating a little on platinum foil.

(*b*) Add to a portion of the diluted acid, a solution of chloride of barium: if sulphuric acid is present, a white precipitate, insoluble in nitric acid, is thrown down (403).

(*c*) Boil a little of the acid with a very small quantity of peroxide of lead ($Pb_3O_4$): if the latter becomes white (owing to its conversion into sulphate of lead), sulphurous acid is present. $Pb_3O_4+SO_2=PbO,SO_3+2PbO$.

(*d*) Nitrate of silver, added to the diluted acid, gives a white curdy precipitate, which is insoluble in nitric acid, if any hydrochloric acid is present (429).

(*e*) Boil a little of the acid, tinged with sulphate of indigo: if the color is bleached, it is probably owing to the presence of nitric acid (452).

(*f*) Neutralize a small portion with ammonia, and add hydrosulphuric acid or hydrosulphate of ammonia: if lead or any other metallic matter is present (except the alkalies and alkaline earths), a precipitate is produced (548).

706. Acetic acid is chiefly employed in the laboratory as a solvent, and for the purpose of acidifying solutions, in cases where hydrochloric and nitric acid would act prejudicially.

*Tartaric Acid* $(2HO,C_8H_4O_{10})$.

707. Tartaric acid sometimes contains a trace of lime and sulphuric acid, but is usually sufficiently pure for analytical purposes. The lime may be detected by neutralizing a portion with ammonia, and adding oxalate of ammonia (218); and the sulphuric acid by chloride of barium (403).

708. Tartaric acid is used as a test for potash, with which it forms a sparingly soluble bitartrate (186). Its property of preventing the precipitation of iron and

some other metals by the alkalies (478), is occasionally made available in analysis. It should be kept in a solid state, and a solution made when required, as when kept in solution it soon becomes mouldy; for this purpose, the crystallized acid may be dissolved in about three times its weight of water.

*Ammonia* ($NH_3$).

709. The liquid ammonia of the shops is generally sufficiently pure for most purposes of analysis; it sometimes, however, contains traces of carbonate, sulphate, and muriate of ammonia, and occasionally chloride of calcium. The carbonate is detected by adding lime water (420); the sulphate by supersaturating with dilute nitric or hydrochloric acid, and testing with chloride of barium (403); the muriate of ammonia may be detected by supersaturating with nitric acid, and adding nitrate of silver (429); and the lime (chloride of calcium) with oxalate of ammonia (218).

Ammonia is used chiefly for the purpose of neutralizing acid solutions, and for precipitating metallic oxides from their solutions, most of which are decomposed by it.

*Hydrosulphate of Ammonia* ($NH_4S,HS$).

710. Hydrosulphate of ammonia is prepared by passing a stream of hydrosulphuric acid gas (699) through a solution of ammonia until it is saturated. To ascertain whether the saturation is complete, a few drops may be tested with sulphate of magnesia; if the ammonia is saturated, this gives no precipitate; but if any free ammonia is left, it throws down the hydrate of magnesia. When first prepared, the solution is almost colorless, but it gradually becomes yellow, owing to partial decomposition, the oxygen of the air combining with the hydrogen, while sulphur is set free, and remains dissolved: when this decomposition has taken place, the addition of an acid causes not only the evolution of hydrosulphuric acid, but also precipitates the dissolved sulphur (440).

711. Hydrosulphate of ammonia is much used, both

in qualitative and quantitative analysis, chiefly for the purpose of precipitating certain metals from their solutions, and for separating the metals of the third class from the alkalies and alkaline earths (593).

*Carbonate of Ammonia* ($2NH_4O,3CO_2$).

712. The common carbonate of ammonia is a sesquicarbonate, or a compound of the neutral carbonate ($NH_4O,CO_2$) and the bicarbonate ($NH_4O,2CO_2$). When the neutral carbonate is required, and it is the best suited for most purposes of analysis, it may be prepared in solution by dissolving one part, by weight, of the crystallized sesquicarbonate, in three or four parts of water, and adding one part of liquid ammonia (sp. gr. 0·96). It is frequently employed in analysis, to precipitate some of the metals as carbonates: it is also used to neutralize acid solutions, and for other purposes.

713. It is occasionally contaminated with traces of animal oil, and sulphate and muriate of ammonia.

(*a*) Heat a small fragment on platinum foil: if any fixed saline impurity is present, it will be left after ignition; and if any charring takes place, it indicates the presence of animal matter.

(*b*) Supersaturate a little of the solution with nitric acid, and add to one portion a few drops of chloride of barium: a white precipitate, insoluble in nitric acid, indicates sulphuric acid (403).

(*c*) To the other portion of the acid solution, add nitrate of silver: if any muriate of ammonia is present, it will cause a white curdy precipitate (429).

*Oxalate of Ammonia* ($NH_4O,C_2O_3+Aq$).

714. This salt, as met with in the shops, is sufficiently pure for all purposes of analysis. Like oxalic acid, it is employed chiefly for the purpose of precipitating lime from its solutions (218); for this purpose, it may be dissolved in about six times its weight of water.

*Phosphate of Soda and Ammonia (Microcosmic Salt).*
($NaO,NH_4O,HO,PO_5+8Aq$.)

715. This salt occasionally contains traces of chloride

of sodium, which may readily be detected by adding a few drops of nitrate of silver to a solution of the salt, acidified with nitric acid, when a curdy white precipitate indicates the presence of the chloride (429).

Microcosmic salt is used almost exclusively in blow-pipe experiments; when heated, it is decomposed, the ammonia and water are expelled, and soda, with excess of phosphoric acid, is left.

### *Potash* (*KO*).

716. On account of its strong affinity for many substances, and its property of readily decomposing others, caustic potash is rarely found free from impurities. Those most commonly met with are organic matter, sulphate and carbonate of potash, chloride of potassium, silicic acid, and alumina.

(*a*) If organic matter is present, the solution of potash is more or less brown, and, on evaporation, leaves a brown residue.

(*b*) Sulphuric acid is detected by diluting the potash with water, supersaturating with nitric or hydrochloric acid, and adding chloride of barium, when, if it is present, the white insoluble sulphate is thrown down (403).

(*c*) If carbonic acid is present, lime water causes a white precipitate, which is soluble with effervescence when the solution is supersaturated with hydrochloric acid (420).

(*d*) A little of the diluted solution is supersaturated with nitric acid, and tested with nitrate of silver: a white curdy precipitate, soluble in ammonia, indicates chlorine or chloride of potassium (429).

(*e*) Neutralize a small portion with hydrochloric acid, and evaporate to dryness: if the residue is not wholly soluble in hydrochloric acid, silica is probably present (425).

(*f*) If alumina is present, it will be precipitated when the potash solution is neutralized with hydrochloric acid, and treated with a *slight* excess of ammonia (241).

717. Potash is used chiefly for the purpose of precipitating some of the metallic oxides from their saline

solutions, which it does on account of its strong affinity for the acids with which they were in combination.

$$CuO,SO_3+KO,HO=KO,SO_3+CuO,HO.$$

It is employed also for neutralizing acid solutions, decomposing organic compounds, and many other purposes. A solution of potash having a specific gravity of about 1060, is a convenient strength for most analytical purposes.

*Carbonate of Potash* ($KO,CO_2+2Aq$).

718. This salt generally contains traces of sulphate and chloride, and occasionally silica and alumina.

(*a*) A solution, supersaturated with nitric acid, and tested with chloride of barium, gives a white precipitate if any sulphuric acid is present (403).

(*b*) A solution similarly acidified, gives, with nitrate of silver, a white curdy precipitate, if it contains chloride of potassium (429).

(*c*) Neutralize a portion of the solution with hydrochloric acid, and evaporate to dryness: if the residue does not wholly dissolve when treated with hydrochloric acid, silica is probably present (425).

(*d*) If carbonate of ammonia causes, in a neutralized solution, a white gelatinous precipitate, alumina is probably present (243).

Carbonate of potash is frequently employed to precipitate metallic oxides and carbonates from their soluble combinations, and for the purpose of neutralizing acid solutions.

*Nitrate of Potash* ($KO,NO_5$).

719. Nitrate of potash often contains traces of sulphate and chloride, and occasionally nitrates of soda and lime.

(*a* and *b*) The sulphate and chloride may be detected with chloride of barium and nitrate of silver (718, *a* and *b*).

(*c*) If lime is present, it causes a precipitate when the solution is treated with oxalate of ammonia (218).

(*d*) The presence of nitrate of soda causes the salt to deliquesce in a moist atmosphere.

720. It is used almost exclusively in the dry state, for the purpose of oxidizing substances, which resist other methods of oxidation; this property is owing to the oxygen of the nitric acid being loosely combined, and at a high temperature readily yielded up to any substance which has a strong affinity for it, such as sulphides, organic matters, &c.

*Iodide of Potassium* (KI).

721. Iodide of potassium is often adulterated with carbonate of potash; sulphate of potash and chloride of potash are also often present. It should always be in the form of well-defined (cubical) crystals, as the adulterated varieties are readily distinguishable by their imperfect crystalline form.

(*a*) Add a little dilute hydrochloric acid: if effervescence takes place, some carbonate is present (419).

(*b*) If sulphates are present, they may be detected by adding chloride of barium, which will, in that case, cause a white precipitate, insoluble in nitric acid (403).

(*c*) Add a little nitrate of silver; this will cause a pale yellow precipitate of iodide of silver, together with chloride of silver, in case any soluble chloride is present. To separate them, filter the mixture, and after washing the precipitate, treat it with a slight excess of ammonia, which dissolves the chloride (if any), and leaves the iodide undissolved (433); on neutralizing the ammoniacal solution with nitric acid, the appearance of a white curdy precipitate indicates the presence of a chloride (429).

Iodide of potassium is employed chiefly as a test for lead, mercury, and occasionally some of the other metals. For use as a reagent, one part of the salt may be dissolved in ten parts of water.

*Chromate of Potash* ($KO,CrO_3$).

722. This salt occasionally contains traces of sulphate of potash, which is readily detected by precipitating a little of the solution with nitrate of baryta, and adding an excess of nitric acid, which redissolves the chromate of baryta, while any sulphate remains insoluble.

It is employed as a test for several of the metallic oxides, with many of which it forms insoluble salts (chromates), of characteristic colors, as the chromate of lead (363), which is bright yellow. For use as a reagent it may be dissolved in ten times its weight of water.

*Cyanide of Potassium* (KCy).

723. Cyanide of potassium is sometimes used in blowpipe experiments, and also as a liquid test. It should be colorless, and entirely soluble in water.

*Ferrocyanide of Potassium* ($K_2$,$FeCy_3$+3Aq). (*Yellow Prussiate of Potash.*)

724. This salt, as met with in commerce, is sufficiently pure for the purposes of testing. It is employed as a test for the persalts of iron, with which it forms a deep blue precipitate of sesquiferrocyanide of iron, or Prussian blue (282). It gives characteristic precipitates, also, with some other metals. For use as a reagent, one part of the salt may be dissolved in fifteen or twenty parts of water.

*Ferridcyanide of Potassium* ($K_3$,$Fe_2Cy_6$). (*Red Prussiate of Potash.*)

725. It occasionally contains traces of the yellow prussiate, which is easily detected by the solution giving a blue precipitate with perchloride of iron (282). It is used as a test for the protosalts of iron, with which it forms a blue precipitate of ferridcyanide of iron (276), which is similar in appearance to that formed by ferrocyanide of potassium with the persalts. It may be dissolved in ten or fifteen parts of water.

*Antimoniate of Potash* (KO,$SbO_5$).

726. This substance seldom or never contains any impurity that can interfere with its action as a test for soda, which is the only use to which it is applied in the laboratory. It must be kept in a well stoppered bottle, and carefully excluded from the air, as the carbonic acid is liable to decompose it, and cause a precipitation of antimonic acid.

*Carbonate of Soda* ($NaO,CO_2+10Aq$).

727. The best method of preparing pure carbonate of soda, is to ignite the crystallized bicarbonate ($NaO,HO,2CO_2$), when the second equivalent of carbonic acid and the water are expelled, and pure anhydrous carbonate is left. The salt of commerce frequently contains a little sulphite and chloride, which may be detected in the manner already detailed (718 *a* and *b*). The more impure varieties contain also traces of sulphide of sodium, and sulphite and hyposulphite of soda. These may be detected by adding dilute sulphuric acid, and passing the evolved gas into a solution of acetate of lead; this should cause a white precipitate of carbonate of lead (422), and not a brown one (438); and no precipitation of sulphur should take place on the addition of the acid.

728. It is employed for the same purposes as carbonate of potash (718); also as a flux for the blowpipe, and for fusing with insoluble silicates, &c. For use as a liquid reagent, one part of the salt may be dissolved in ten parts of water.

*Phosphate of Soda* ($2NaO,HO,PO_5+24Aq$).

729. This salt sometimes contains a little sulphate and chloride. To detect these impurities, add to one portion, in solution, chloride of barium, and to the other nitrate of silver, and supersaturate both with nitric acid: if the precipitate does not entirely dissolve in either case, a sulphate or chloride is present (403, 429).

It is employed chiefly as a test for magnesia, with which it forms, in the presence of ammoniacal salts, the double phosphate of magnesia and ammonia (206). For the purposes of testing, it may be dissolved in ten parts of water.

*Borax (Biborate of Soda)*, ($NaO,2BO_3+10Aq$).

730. Borax occasionally contains traces of sulphate and chloride, which may be detected in the same way as in the phosphate of soda (729). It is employed almost exclusively as a flux in blowpipe experiments,

for which purpose it is admirably adapted; the second equivalent of boracic acid which it contains, exerts a strong affinity for bases at a high temperature, and is capable of displacing several acids from their combinations; it also forms many double compounds and mixtures which are readily fusible.

*Lime Water (CaO in Water).*

731. This reagent is prepared by digesting hydrate of lime (CaO,HO) in cold distilled water for an hour or two, stirring the mixture occasionally, and when the undissolved portion of the lime has subsided, pouring off the clear solution, and filtering if necessary. As it is liable to spoil when exposed to the air, owing to the absorption of carbonic acid, it should be kept in a well-stoppered bottle.

732. Lime water should be sufficiently strong to turn the yellow color of turmeric instantly and decidedly brown; and, when tested with carbonate of soda, should throw down a copious white precipitate of carbonate of lime (214). It is used as a test for carbonic acid and some of the organic acids; for expelling ammonia from its combinations, and for many other purposes.

*Sulphate of Lime* ($CaO,SO_3+2Aq$).

733. Sulphate of lime being very sparingly soluble in water, is always used in the form of a saturated solution, which is prepared by digesting the sulphate in water, stirring it occasionally, and pouring off the clear solution from the undissolved portion. It is used chiefly as a test for some of the organic acids, and for distinguishing baryta from strontia. The solution ought to give an immediate precipitate of sulphate of baryta, when tested with chloride of barium (225).

*Chloride of Calcium* (CaCl).

734. This substance occasionally contains a little free acid, and traces of iron. The first is detected by test paper, and the latter, if present, causes hydrosulphate of ammonia to throw down in the solution a black pre-

cipitate, or to impart a greenish tint to the liquid (279). As a reagent, chloride of calcium is employed chiefly in testing for some of the organic acids. It is also of great use in the laboratory as a drying agent, having so strong an affinity for water, that a moist gas passed over it, is rapidly and completely deprived of its water. For this purpose the chloride need not be absolutely pure: it should not be fused, but merely dried, as the unfused is more porous, and consequently offers a larger amount of surface to any gas passed over it.

*Chloride of Barium* (BaCl+2Aq).

735. Chloride of barium sometimes contains traces of iron and lime. It should not be discolored by hydrosulphate of ammonia (279), and, after being treated with a slight excess of sulphuric acid, and filtered, the clear solution should leave no fixed residue when evaporated on platinum foil; because the whole of the baryta is separated by the sulphuric acid, and any other fixed matter must be some impurity.

It is used chiefly for the purpose of testing for acids (558), especially sulphuric, with which it forms the insoluble sulphate of baryta (403). For use, one part of the salt may be dissolved in ten parts of water.

*Nitrate of Baryta* ($BaO,NO_5$).

736. Nitrate of baryta is liable to the same impurities as chloride of barium (735), and they may be detected in the same way. It should also be free from any chloride, which may be known by adding nitrate of silver (429). Its uses are the same as those of chloride of barium, for which it is occasionally substituted in cases when the addition of the chloride would interfere with the subsequent stages of an analysis, as when we have to test for chlorides in the same solution (605). For use, it may be dissolved in ten parts of water.

*Perchloride of Iron* ($Fe_2Cl_3$).

737. This salt is liable to contain a little free acid, and traces of the protochloride (FeCl). The free acid

is detected in the manner described in (535, *b*); and if any protosalt of iron is present, the solution gives a blue color with ferridcyanide of potassium (276). It is used as a test for some of the organic acids, and is also sometimes useful in the determination of phosphoric acid. It may be dissolved in five parts of water.

*Nitrate of Cobalt* ($CoO,NO_5+6Aq$).

738. This reagent is used chiefly for the detection of alumina, zinc, magnesia, and some other substances, by means of the blowpipe (124). The solution employed for this purpose may contain one part of the salt dissolved in ten of water.

*Sulphate of Copper* ($CuO,SO_3+5Aq$).

739. This salt is occasionally used as a test for arsenic (311), and for other purposes: it may be dissolved in ten parts of water. The *ammonio-sulphate of copper* ($CuO,2NH_3,HO,SO_3$), which is also used in testing for arsenic, is prepared by adding ammonia to the solution of sulphate of copper, until the precipitate at first formed is nearly all dissolved, when the solution is filtered, and kept for use.

*Acetate of Lead* ($PbO,C_4H_3O_3+3Aq$).

740. Acetate of lead is used as a test for several acids, which form with oxide of lead insoluble salts. For testing, one part of the salt may be dissolved in ten parts of water.

*Subacetate of Lead* ($3PbO,C_4H_3O_3$).

741. The subacetate is prepared by boiling together equal weights of the neutral acetate (740) and protoxide of lead (PbO) in water, and filtering the solution, which must be kept in a well-stoppered bottle, as it is easily decomposed when in contact with the air, owing to the strong affinity of the oxide of lead for carbonic acid. Both this and the neutral acetate are used in testing for hydrosulphuric acid, and for some of the other acids, especially carbonic.

*Nitrate of Silver* ($AgO,NO_5$).

742. This reagent is sometimes adulterated with nitrate of potash, and occasionally contains traces of copper and lead. When precipitated by a slight excess of hydrochloric acid, the filtered solution ought to leave no fixed residue when evaporated on platinum foil, as the whole of the silver would be thrown down (377), and any impurity would remain in solution. Copper is detected by adding ammonia in excess to the solution, when it will give the liquid a blue tinge (369). Nitrate of silver is used chiefly as a test for chlorine (chlorides and hydrochloric acid), and also for phosphoric and some of the other acids. For use as a reagent, one part of the salt may be dissolved in twenty parts of water.

743. The *ammonio-nitrate of silver* ($AgO,2NH_3,NO_5$), used as a test for arsenic, is prepared by adding ammonia to a solution of the nitrate, until the precipitate at first thrown down is nearly all redissolved, and filtering from the undissolved oxide.

*Perchloride of Mercury* ($HgCl_2$).

744. This is occasionally employed as a test for hydriodic and some other acids, and also for some kinds of organic matter: for this purpose it may be dissolved in twenty parts of water.

*Protochloride of Tin* ($SnCl$).

745. Protochloride of tin is prepared by boiling metallic tin in strong hydrochloric acid, care being taken that a portion of the metal remains undissolved, as otherwise a little perchloride might be formed; the solution is then filtered, acidified with a few drops of hydrochloric acid, and diluted with about four times its bulk of water. A few fragments of metallic tin should be kept in the solution, in order to prevent the formation of any perchloride.

746. Protochloride of tin is employed chiefly as a test for gold and mercury, and also as a deoxidizing agent, for which purpose it is well adapted, on account of its strong tendency to combine with oxygen or chlorine.

It occasionally contains traces of lead and iron, which may be detected by adding hydrosulphate of ammonia in excess to the solution, when, if pure, the precipitate is wholly redissolved, but, if either of those metals is present, a black residue is left, since their sulphides are insoluble in the hydrosulphate.

*Perchloride of Gold* ($AuCl_3$).

747. This reagent is used almost exclusively as a test for the protosalts of tin (386), so that a very small quantity will be found sufficient for the purpose of testing. One part of the salt may be dissolved in thirty parts of water.

*Bichloride of Platinum* ($PtCl_2$).

748. Bichloride of platinum is employed only as a test for potash, soda, and ammonia; it may be dissolved in about ten parts of alcohol.

*Sulphate of Indigo.*

749. This substance may be prepared in solution, by dissolving a little indigo in strong sulphuric acid, and diluting the acid solution with water, so as to form a pale blue liquid. It is used chiefly as a test for nitric acid and chlorine, by which it is decomposed, and its color discharged.

*Solution of Starch* ($C_{12}H_{10}O_{10}$).

750. This is made by gently boiling starch with water. It is employed as a test for iodine, for which purpose small pieces of thread on paper may be steeped in the solution, dried, and kept for use.

*Black Flux.*

751. Black flux is an intimate mixture of carbonate of potash and finely divided charcoal, and is prepared by deflagrating in an iron spoon or crucible, a mixture of two parts of bitartrate of potash and one of nitre. It is used as a reducing flux in blowpipe experiments.

### *Distilled Water* (*HO*).

752. Pure distilled water is prepared by carefully distilling any of the common kinds of water either in a still or retort, rejecting the first and last portions (62). For many purposes, rain water, when collected at a distance from towns or manufactories, and boiled and filtered, will be found sufficiently pure; but in analytical experiments, distilled water ought always to be used.

753. Before taking it into use, it should be tested with the following reagents:—

(*a*) Litmus and turmeric paper, for free acids and alkalies.

(*b*) Chloride of barium for sulphates (403).

(*c*) Nitrate of silver for chlorides (429). The mixture shortly becomes dark-colored, especially if organic matter is present.

(*d*) Oxalate of ammonia for lime (218).

(*e*) Lime water for carbonic acid (420).

(*f*) Hydrosulphate of ammonia for any metals of the third or fourth class.

(*g*) When heated on platinum foil, it should leave no trace of solid residue.

Distilled water is used chiefly as a solvent, and for washing precipitates, besides many other purposes to which it is constantly applied.

### *Alcohol* ($C_4H_5,O,HO$).

754. The alcohol commonly used in chemical experiments should have a specific gravity of about 0·83, except in cases where absolute alcohol is required, when it should be 0·796. When evaporated on platinum foil, it should leave no residue, and should not change the color of litmus paper. It is used chiefly as a solvent, and for the purpose of facilitating the precipitation of substances which are less soluble in it than in water.

# APPENDIX.

---

## WEIGHTS AND MEASURES.

### *Troy or Apothecaries' Weight.*

| Pound. | | Ounces. | | Drachms. | | Scruples. | | Grains. | | French Grammes. |
|---|---|---|---|---|---|---|---|---|---|---|
| 1 | = | 12 | = | 96 | = | 288 | = | 5760 | = | 372·96 |
| | | 1 | = | 8 | = | 24 | = | 480 | = | 31·08 |
| | | | | 1 | = | 3 | = | 60 | = | 3·885 |
| | | | | | | 1 | = | 20 | = | 1·295 |
| | | | | | | | | 1 | = | 0·0647 |

### *Avoirdupois Weight.*

| Pound. | | Ounces. | | Drachms. | | Grains. | | French Grammes |
|---|---|---|---|---|---|---|---|---|
| 1 | = | 16 | = | 256 | = | 7000 | = | 453·25 |
| | | 1 | = | 16 | = | 437·5 | = | 28·328 |
| | | | | 1 | = | 27·343 | = | 1·77 |

### *Imperial Measure.*

| Gallon. | | Pints. | | Fluid Ounces. | | Fluid Drachms. | | Minims. |
|---|---|---|---|---|---|---|---|---|
| 1 | = | 8 | = | 160 | = | 1280 | = | 76,800 |
| | | 1 | = | 20 | = | 160 | = | 9,600 |
| | | | | 1 | = | 8 | = | 480 |
| | | | | | | 1 | = | 60 |

### *Weight of Water at 62°, contained in the Imperial Gallon, &c.*

| | | | Grains. |
|---|---|---|---|
| 1 Imperial Gallon | . . | = . . | 70,000 |
| 1 " Pint | . . | = . . | 8,750 |
| 1 " Fluid Ounce | . | = . . | 437·5 |
| 1 " Fluid Drachm | . | = . . | 54·7 |
| 1 " Minim | . . | = . . | 0·91 |

*Cubic Inches contained in the Imperial Gallon, &c.*

| | | | Cubic inches. |
|---|---|---|---|
| 1 Imperial Gallon | . . | = . . . | 277·273 |
| 1 " Pint | . . | = . . . | 34·659 |
| 1 " Fluid Ounce | . | = . . . | 1·732 |
| 1 " Fluid Drachm | . | = . . . | 0·2166 |
| 1 " Minim | . . | = . . . | 0·0036 |

## FRENCH WEIGHTS AND MEASURES.

*Measures of Length.*

| | | English Inches. | | Mil. | Fur. | Yds. | Feet. | In. |
|---|---|---|---|---|---|---|---|---|
| Millimetre | = | ·03937 | | | | | | |
| Centimetre | = | ·39371 | | | | | | |
| Decimetre | = | 3·93710 | | | | | | |
| Metre | = | 39·37100 | | | | | | |
| Decametre | = | 393·71000 | = | 0 | 0 | 10 | 2 | 9·7 |
| Hecatometre | = | 3937·10000 | = | 0 | 0 | 109 | 1 | 1 |
| Kilometre | = | 39371·00000 | = | 0 | 4 | 213 | 4 | 10·2 |
| Myriometre | = | 393710·00000 | = | 6 | 1 | 156 | 0 | 6 |

*Measures of Capacity.*

| | | | | English Imperial Measure. | | | | |
|---|---|---|---|---|---|---|---|---|
| | | Cubic Inches. | | Gall. | Pints. | F.oz. | F.drms. | Min. |
| Millilitre | = | ·06102 | = | 0 | 0 | 0 | 0 | 16·3 |
| Centilitre | = | ·61028 | = | 0 | 0 | 0 | 2 | 42 |
| Decilitre | = | 6·10280 | = | 0 | 0 | 3 | 3 | 2 |
| Litre | = | 61·02800 | = | 0 | 1 | 15 | 1 | 43 |
| Decalitre | = | 610·28000 | = | 2 | 1 | 12 | 1 | 16 |
| Hecatolitre | = | 6102·80000 | = | 22 | 0 | 1 | 4 | 48 |
| Kilolitre | = | 61028·00000 | = | 220 | 0 | 12 | 6 | 24 |
| Myriolitre | = | 610280·00000 | = | 2200 | 7 | 13 | 4 | 48 |

*Measures of Weight.*

| | | | | *Avoirdupois.* | | |
|---|---|---|---|---|---|---|
| | | English Grains. | | Poun. | Oun. | Dram. |
| Milligramme | = | ·0154 | | | | |
| Centigramme | = | ·1544 | | | | |
| Decigramme | = | 1·5444 | | | | |
| Gramme | = | 15·4440 | | | | |
| Decagramme | = | 154·4402 | = | 0 | 0 | 5·65 |
| Hecatogramme | = | 1554·4023 | = | 0 | 3 | 8·5 |
| Kilogramme | = | 15444·0234 | = | 2 | 3 | 5 |
| Myriogramme | = | 154440·2344 | = | 22 | 1 | 2 |

# TABLE I.

*Showing the Quantity of Oil of Vitriol* ($HO,SO_3$) *of sp. gr.* 1·8485, *and of Anhydrous Acid* ($SO_3$), *in* 100 *Parts of dilute Sulphuric Acid, of different Specific Gravities* (*Ure*).

| Liquid Acid. | Sp. Gr. | Dry Acid. | Liquid Acid. | Sp. Gr. | Dry Acid. |
|---|---|---|---|---|---|
| 100 | 1·8485 | 81·54 | 65 | 1·5390 | 53·00 |
| 99 | 1·8475 | 88·72 | 64 | 1·5280 | 52·18 |
| 98 | 1·8460 | 79·90 | 63 | 1·5170 | 51·37 |
| 97 | 1·8439 | 79·09 | 62 | 1·5066 | 50,55 |
| 96 | 1·8410 | 78·28 | 61 | 1·4960 | 49·74 |
| 95 | 1·8376 | 77·46 | 60 | 1·4860 | 48·92 |
| 94 | 1·8336 | 76·65 | 59 | 1·4760 | 48·11 |
| 93 | 1·8290 | 75·83 | 58 | 1·4660 | 47·29 |
| 92 | 1·8233 | 75·02 | 57 | 1·4560 | 46·48 |
| 91 | 1·8179 | 74·20 | 56 | 1·4460 | 45·66 |
| 90 | 1·8115 | 73·39 | 55 | 1·4360 | 44·85 |
| 89 | 1·8043 | 72·57 | 54 | 1·4265 | 44·03 |
| 88 | 1·7962 | 71·75 | 53 | 1·4170 | 43·22 |
| 87 | 1·7870 | 70·94 | 52 | 1·4073 | 42·40 |
| 86 | 1·7774 | 70·12 | 51 | 1·3977 | 41·58 |
| 85 | 1·7673 | 69·31 | 50 | 1·3884 | 40·77 |
| 84 | 1·7570 | 68·49 | 49 | 1·3788 | 39·95 |
| 83 | 1·7465 | 67·68 | 48 | 1·3697 | 39·15 |
| 82 | 1·7360 | 66·86 | 47 | 1·3612 | 38·32 |
| 81 | 1·7245 | 66·05 | 46 | 1·3530 | 37·51 |
| 80 | 1·7120 | 65·23 | 45 | 1·3440 | 36·69 |
| 79 | 1 6993 | 64·42 | 44 | 1·3345 | 35·88 |
| 78 | 1·6870 | 63·60 | 43 | 1·3255 | 35·06 |
| 77 | 1·6750 | 62·78 | 42 | 1·3165 | 34·25 |
| 76 | 1 6630 | 61·97 | 41 | 1·3080 | 33·43 |
| 75 | 1·6520 | 61·15 | 40 | 1 2999 | 32·61 |
| 74 | 1·6415 | 60·34 | 39 | 1·2913 | 31·80 |
| 73 | 1·6321 | 59·82 | 38 | 1·2826 | 30·98 |
| 72 | 1·6204 | 58·71 | 37 | 1·2740 | 30·17 |
| 71 | 1·6090 | 57·89 | 36 | 1·2654 | 29·35 |
| 70 | 1·5975 | 57·08 | 35 | 1·2572 | 28·54 |
| 69 | 1·5868 | 56·26 | 34 | 1·2490 | 27·72 |
| 68 | 1·5760 | 55·45 | 33 | 1·2409 | 26 91 |
| 67 | 1·5648 | 54·63 | 32 | 1·2334 | 26·09 |
| 66 | 1·5503 | 53·82 | 31 | 1·2260 | 25·82 |

TABLE I.—*Continued.*

| Liquid Acid. | Sp. Gr. | Dry Acid. | Liquid Acid. | Sp. Gr. | Dry Acid. |
|---|---|---|---|---|---|
| 30 | 1·2184 | 24·46 | 15 | 1·1019 | 12·23 |
| 29 | 1·2108 | 23·65 | 14 | 1·0953 | 11·41 |
| 28 | 1·2032 | 22·83 | 13 | 1·0887 | 10·60 |
| 27 | 1·1956 | 22·01 | 12 | 1·0809 | 9·78 |
| 26 | 1·1876 | 21·20 | 11 | 1·0743 | 8·97 |
| 25 | 1·1792 | 20·38 | 10 | 1·0682 | 8·15 |
| 24 | 1·1706 | 19·57 | 9 | 1·0614 | 7·34 |
| 23 | 1·1626 | 18·75 | 8 | 1·0544 | 6·52 |
| 22 | 1·1549 | 17·94 | 7 | 1·0477 | 5·71 |
| 21 | 1·1480 | 17·12 | 6 | 1·0405 | 4·89 |
| 20 | 1·1410 | 16·31 | 5 | 1·0336 | 4·08 |
| 19 | 1·1330 | 15·49 | 4 | 1·0268 | 3·26 |
| 18 | 1·1246 | 14·68 | 3 | 1·0206 | 2·446 |
| 17 | 1·1165 | 13 84 | 2 | 1·0140 | 1·63 |
| 16 | 1·1090 | 13·05 | 1 | 1·0074 | 0·8154 |

TABLE II.

*Showing the Quantity of Real or Anhydrous Nitric Acid* ($NO_5$) *in* 100 *Parts of Liquid Acid, of different Specific Gravities* (*Ure*).

| Specific Gravity. | Real acid in 100 parts of the Liquid. | Specific Gravity. | Real acid in 100 parts of the Liquid. | Specific Gravity. | Real acid in 100 parts of the Liquid. |
|---|---|---|---|---|---|
| 1·5000 | 79·700 | 1·4600 | 68·542 | 1·4065 | 57·384 |
| 1·4980 | 78·903 | 1·4570 | 67·745 | 1·4023 | 56·587 |
| 1·4960 | 78·106 | 1·4530 | 66·948 | 1·3978 | 55·790 |
| 1·4940 | 77·309 | 1·4500 | 66·155 | 1·3945 | 54 993 |
| 1 4910 | 76·512 | 1·4460 | 65·354 | 1·3882 | 54·196 |
| 1·4880 | 75·715 | 1·4424 | 64 557 | 1·3833 | 53·399 |
| 1·4850 | 74·918 | 1·4385 | 63·760 | 1 3783 | 52·602 |
| 1·4820 | 74·121 | 1·4346 | 62·963 | 1·3732 | 51 805 |
| 1·4790 | 73·324 | 1·4306 | 62·166 | 1·3681 | 51·068 |
| 1·4760 | 72·527 | 1·4269 | 61·369 | 1·3630 | 50 211 |
| 1·4730 | 71·730 | 1·4228 | 60·572 | 1·3579 | 49·414 |
| 1·4700 | 70·933 | 1·4189 | 59·775 | 1·3529 | 48·617 |
| 1·4670 | 70·136 | 1·4147 | 58·978 | 1·3477 | 47·820 |
| 1·4640 | 69·339 | 1·4107 | 58·181 | 1·3427 | 47·023 |

TABLE II.—*Continued.*

| Specific Gravity. | Real acid in 100 parts of the Liquid. | Specific Gravity. | Real acid in 100 parts of the Liquid. | Specific Gravity. | Real acid in 100 parts of the Liquid. |
|---|---|---|---|---|---|
| 1·3376 | 46·226 | 1·2212 | 30·286 | 1·1051 | 15·143 |
| 1·3323 | 45·429 | 1·2148 | 29·489 | 1·0993 | 14·346 |
| 1·3270 | 44·632 | 1·2084 | 28·692 | 1·0935 | 13·549 |
| 1·3216 | 43·835 | 1·2019 | 27·895 | 1·0878 | 12·752 |
| 1·3163 | 43·038 | 1·1958 | 27·098 | 1·0821 | 11·955 |
| 1·3110 | 42·241 | 1·1895 | 26·301 | 1·0764 | 11·158 |
| 1·3056 | 41·444 | 1·1833 | 25·504 | 1·0708 | 10·361 |
| 1·3001 | 40·647 | 1·1770 | 24·707 | 1·0651 | 9·564 |
| 1·2947 | 39·850 | 1·1709 | 23·910 | 1·0595 | 8·767 |
| 1·2887 | 39·053 | 1·1648 | 23·113 | 1·0540 | 7·970 |
| 1·2826 | 38·256 | 1·1587 | 22·316 | 1·0485 | 7·173 |
| 1·2765 | 37·459 | 1·1526 | 21·519 | 1·0430 | 6·376 |
| 1·2705 | 36·662 | 1·1465 | 20·722 | 1·0375 | 5·579 |
| 1·2644 | 35·865 | 1·1403 | 19·925 | 1·0320 | 4·782 |
| 1·2583 | 35·068 | 1·1345 | 19·128 | 1·0267 | 3·985 |
| 1·2523 | 34·271 | 1·1286 | 18·331 | 1·0212 | 3·188 |
| 1·2462 | 33·474 | 1·1227 | 17·534 | 1·0159 | 2·391 |
| 1·2402 | 32·677 | 1·1168 | 16·737 | 1·0106 | 1·594 |
| 1·2341 | 31·880 | 1·1109 | 15·940 | 1·0053 | 0·797 |
| 1·2277 | 31·083 | | | | |

TABLE III.

*Showing the Quantity of Anhydrous Hydrochloric Acid* (HCl) *in the Liquid Acid of different Specific Gravities* (*Ure*).

| Acid of 120 in 100. | Specific Gravity. | Chlorine. | Hydrochloric Gas. | Acid of 120 in 100. | Specific Gravity. | Chlorine. | Hydrochloric Gas. |
|---|---|---|---|---|---|---|---|
| 100 | 1·2000 | 39·675 | 40·777 | 92 | 1·1857 | 36·503 | 37·516 |
| 99 | 1·1982 | 39·278 | 40·369 | 91 | 1·1846 | 36·107 | 37·108 |
| 98 | 1·1964 | 38·882 | 39·961 | 90 | 1·1822 | 35·707 | 36·700 |
| 97 | 1·1946 | 38·485 | 39·554 | 89 | 1·1802 | 35·310 | 36·292 |
| 96 | 1·1928 | 38·089 | 39·146 | 88 | 1·1782 | 34·913 | 35·884 |
| 95 | 1·1910 | 37·692 | 38·738 | 87 | 1·1762 | 34·517 | 35·476 |
| 94 | 1·1893 | 37·296 | 38·330 | 86 | 1·1741 | 34·121 | 35·068 |
| 93 | 1·1875 | 36·900 | 37·923 | 85 | 1·1721 | 33·724 | 34·660 |

TABLE III.—*Continued.*

| Acid of 120 in 100. | Specific Gravity. | Chlorine. | Hydrochloric Gas. | Acid of 120 in 100. | Specific Gravity. | Chlorine. | Hydrochloric Gas. |
|---|---|---|---|---|---|---|---|
| 84 | 1·1701 | 33·328 | 34·252 | 42 | 1·0838 | 16·664 | 17·126 |
| 83 | 1·1681 | 32·931 | 33·845 | 41 | 1·0818 | 16·267 | 16·718 |
| 82 | 1·1661 | 32·535 | 33·437 | 40 | 1·0798 | 15·870 | 16·310 |
| 81 | 1·1641 | 32·136 | 33·029 | 39 | 1·0778 | 15·474 | 15·902 |
| 80 | 1·1620 | 31·743 | 32·621 | 38 | 1·0758 | 15·077 | 15·494 |
| 79 | 1·1599 | 31·343 | 32·213 | 37 | 1·0738 | 14·680 | 15·087 |
| 78 | 1·1578 | 30·946 | 31·805 | 36 | 1·0718 | 14·284 | 14·679 |
| 77 | 1·1557 | 30·550 | 31·398 | 35 | 1·0697 | 13·887 | 14·271 |
| 76 | 1·1536 | 30·153 | 30·990 | 34 | 1·0677 | 13·490 | 13·863 |
| 75 | 1·1515 | 29·757 | 30·582 | 33 | 1·0657 | 13·094 | 13·456 |
| 74 | 1·1494 | 29·361 | 30·174 | 32 | 1·0637 | 12·597 | 13·049 |
| 73 | 1·1473 | 28·964 | 29·767 | 31 | 1·0617 | 12·300 | 12·641 |
| 72 | 1·1452 | 28·567 | 29·359 | 30 | 1·0597 | 11·903 | 12·233 |
| 71 | 1·1431 | 28·171 | 28·951 | 29 | 1·0577 | 11·506 | 11·825 |
| 70 | 1·1410 | 27·772 | 28 544 | 28 | 1·0557 | 11·109 | 11·418 |
| 69 | 1·1389 | 27·376 | 28·136 | 27 | 1·0537 | 10·712 | 11·010 |
| 68 | 1·1369 | 26·979 | 27·728 | 26 | 1·0517 | 10·316 | 10.602 |
| 67 | 1·1349 | 26·583 | 27·321 | 25 | 1·0497 | 9·919 | 10·194 |
| 66 | 1·1328 | 26·186 | 26·913 | 24 | 1·0477 | 9·522 | 9·786 |
| 65 | 1·1308 | 25·789 | 26·505 | 23 | 1·0457 | 9·126 | 9·379 |
| 64 | 1·1287 | 25·392 | 26·098 | 22 | 1·0437 | 8·729 | 8·971 |
| 63 | 1·1267 | 24·996 | 25·690 | 21 | 1·0417 | 8·332 | 8·563 |
| 62 | 1·1247 | 24·599 | 25·282 | 20 | 1·0397 | 7·935 | 8·155 |
| 61 | 1·1226 | 24·202 | 24·874 | 19 | 1·0377 | 7·538 | 7·747 |
| 60 | 1·1206 | 23·805 | 24·466 | 18 | 1·0357 | 7·141 | 7·340 |
| 59 | 1·1185 | 23·408 | 24·058 | 17 | 1·0337 | 6·745 | 6·932 |
| 58 | 1·1164 | 23·812 | 23·650 | 16 | 1·0318 | 6·348 | 6·524 |
| 57 | 1·1143 | 22·615 | 23·242 | 15 | 1·0298 | 5·951 | 6·116 |
| 56 | 1·1123 | 22·218 | 22·834 | 14 | 1·0279 | 5·554 | 5·709 |
| 55 | 1·1102 | 21·822 | 22·426 | 13 | 1·0259 | 5·158 | 5·301 |
| 54 | 1·1082 | 21·425 | 22·019 | 12 | 1·0239 | 4·762 | 4·893 |
| 53 | 1·1061 | 21·028 | 21·611 | 11 | 1·0220 | 4·365 | 4·486 |
| 52 | 1·1041 | 20·632 | 21·203 | 10 | 1·0200 | 3·998 | 4 078 |
| 51 | 1·1020 | 20·235 | 20·796 | 9 | 1·0180 | 3·571 | 3 670 |
| 50 | 1·1000 | 19·837 | 20·388 | 8 | 1·0160 | 3·174 | 3·262 |
| 49 | 1·0980 | 19·440 | 19·980 | 7 | 1·0140 | 2·778 | 2·854 |
| 48 | 1·0960 | 19 044 | 19·572 | 6 | 1·0120 | 2·381 | 2·447 |
| 47 | 1·0939 | 18·647 | 19·165 | 5 | 1·0110 | 1·984 | 2·039 |
| 46 | 1·0919 | 18·250 | 18·757 | 4 | 1·0080 | 1·588 | 1·631 |
| 45 | 1·0899 | 17·854 | 18·349 | 3 | 1·0060 | 1·191 | 1·224 |
| 44 | 1·0879 | 17·427 | 17·941 | 2 | 1·0040 | 0·795 | 0·816 |
| 43 | 1·0859 | 17·060 | 17·534 | 1 | 1·0020 | 0·397 | 0·408 |

## TABLE IV.

*Showing the Quantity of Anhydrous Potash (KO) in Solutions of different Specific Gravities (Dalton).*

| Specific Gravity. | Potash per cent. | Boiling-point. | Specific Gravity. | Potash per cent. | Boiling-point. |
|---|---|---|---|---|---|
| 1·63 | 51·2 | 329° | 1·33 | 26·3 | 229° |
| 1·60 | 46·7 | 290 | 1·28 | 23·4 | 224 |
| 1·52 | 42·9 | 276 | 1·23 | 19 5 | 220 |
| 1·47 | 39·6 | 265 | 1·19 | 16·2 | 218 |
| 1·44 | 36·8 | 255 | 1·15 | 13· | 215 |
| 1·42 | 34·4 | 246 | 1·11 | 9·5 | 214 |
| 1·39 | 32·4 | 240 | 1·06 | 4·7 | 213 |
| 1·36 | 29·4 | 234 | | | |

## TABLE V.

*Showing the Quantity of Anhydrous Soda (NaO) in solutions of different Specific Gravities (Dalton).*

| Specific Gravity. | Soda per cent. | Boiling-point. | Specific Gravity. | Soda per cent. | Boiling-point. |
|---|---|---|---|---|---|
| 2 00 | 77·8 | —°? | 1·40 | 29·0 | 242° |
| 1·85 | 63·6 | 600 | 1·36 | 26·0 | 235 |
| 1·72 | 53·8 | 400 | 1·32 | 22·0 | 228 |
| 1·63 | 46·6 | 300 | 1·29 | 19·0 | 224 |
| 1 56 | 41·2 | 280 | 1·23 | 16·0 | 220 |
| 1·50 | 36·8 | 265 | 1·18 | 13·0 | 217 |
| 1·47 | 34·0 | 255 | 1·12 | 9·8 | 214 |
| 1·44 | 31 0 | 248 | 1·06 | 4·7 | 213 |

## TABLE VI.

*Showing the Quantity of Ammoniacal Gas* ($NH_3$) *in Aqueous Solutions of different Specific Gravities (Dalton).*

| Specific Gravity. | Grains of Ammonia in 100 grains of the liquid. | Boiling-points. | Volumes of gas in one volume of the liquid. |
|---|---|---|---|
| ·850 | 35 3 | 26° | 494 |
| ·860 | 32 6 | 38 | 456 |
| ·870 | 29·9 | 50 | 419 |
| ·880 | 27·3 | 62 | 382 |
| ·890 | 24 7 | 74 | 346 |
| ·900 | 22·2 | 86 | 311 |
| ·910 | 19·8 | 98 | 277 |
| ·920 | 17·4 | 110 | 244 |
| ·930 | 15·1 | 122 | 211 |
| ·940 | 12·8 | 134 | 180 |
| ·950 | 10 5 | 146 | 147 |
| ·960 | 8·3 | 158 | 116 |
| ·970 | 6·2 | 173 | 87 |
| ·980 | 4·4 | 187 | 58 |
| ·990 | 2·0 | 196 | 28 |

## TABLE VII.

*Showing the Quantity of Absolute Alcohol* ($C_4H_5O,HO$) *contained in Diluted Alcohol of different Specific Gravities (Fownes).*

| Sp. Gr. at 60°. | Percentage of real Alcohol. | Sp. Gr. at 60°. | Percentage of real Alcohol. | Sp. Gr. at 60°. | Percentage of real Alcohol. |
|---|---|---|---|---|---|
| 0·9991 | 0 5 | 0·9841 | 10 | 0·9716 | 20 |
| 0·9981 | 1 | 0·9828 | 11 | 0·9704 | 21 |
| 0·9965 | 2 | 0·9815 | 12 | 0·9691 | 22 |
| 0 9947 | 3 | 0·9802 | 13 | 0·9678 | 23 |
| 0 9930 | 4 | 0·9789 | 14 | 0·9665 | 24 |
| 0·9914 | 5 | 0·9778 | 15 | 0·9652 | 25 |
| 0 9898 | 6 | 0·9766 | 16 | 0·9638 | 26 |
| 0·9884 | 7 | 0 9753 | 17 | 0 9623 | 27 |
| 0 9869 | 8 | 0·9741 | 18 | 0·9609 | 28 |
| 0·9855 | 9 | 0·9728 | 19 | 0·9593 | 29 |

TABLE VII.—*Continued.*

| Sp. Gr. at 60°. | Percentage of real Alcohol. | Sp. Gr. at 60°. | Percentage of real Alcohol. | Sp. Gr. at 60°. | Percentage of real Alcohol. |
|---|---|---|---|---|---|
| 0·9578 | 30 | 0·9090 | 54 | 0·8533 | 78 |
| 0·9560 | 31 | 0·9069 | 55 | 0·8508 | 79 |
| 0·9544 | 32 | 0·9047 | 56 | 0·8483 | 80 |
| 0·9528 | 33 | 0·9025 | 57 | 0·8459 | 81 |
| 0·9511 | 34 | 0·9001 | 58 | 0·8434 | 82 |
| 0·9490 | 35 | 0·8979 | 59 | 0·8408 | 83 |
| 0·9470 | 36 | 0·8956 | 60 | 0·8382 | 84 |
| 0·9452 | 37 | 0·8932 | 61 | 0·8357 | 85 |
| 0·9434 | 38 | 0·8908 | 62 | 0·8331 | 86 |
| 0·9416 | 39 | 0·8886 | 63 | 0·8305 | 87 |
| 0·9396 | 40 | 0·8863 | 64 | 0·8279 | 88 |
| 0·9376 | 41 | 0·8840 | 65 | 0·8254 | 89 |
| 0·9356 | 42 | 0·8816 | 66 | 0·8228 | 90 |
| 0·9335 | 43 | 0·8793 | 67 | 0·8199 | 91 |
| 0·9314 | 44 | 0·8769 | 68 | 0·8172 | 92 |
| 0·9292 | 45 | 0·8745 | 69 | 0·8145 | 93 |
| 0·9270 | 46 | 0·8721 | 70 | 0·8118 | 94 |
| 0·9249 | 47 | 0·8696 | 71 | 0·8089 | 95 |
| 0·9228 | 48 | 0·8672 | 72 | 0·8061 | 96 |
| 0·9206 | 49 | 0·8649 | 73 | 0·8031 | 97 |
| 0·9184 | 50 | 0·8625 | 74 | 0·8001 | 98 |
| 0·9160 | 51 | 0·8603 | 75 | 0·7969 | 99 |
| 0·9135 | 52 | 0·8581 | 76 | 0·7938 | 100 |
| 0·9113 | 53 | 0·8557 | 77 | | |

## TABLE VIII.

*Showing the Specific Gravities of mixtures of Ether and Alcohol in different proportions (Dalton).*

| Specific Gravity. | Ether. | Alcohol (sp. gr. 830). | Specific Gravity. | Ether. | Alcohol (sp. gr. 830). |
|---|---|---|---|---|---|
| 724 | 100 | 0 | 792 | 40 | 60 |
| 732 | 90 | 10 | 804 | 30 | 70 |
| 744 | 80 | 20 | 816 | 20 | 80 |
| 756 | 70 | 30 | 828 | 10 | 90 |
| 768 | 60 | 40 | 830 | 0 | 100 |
| 780 | 50 | 50 | | | |

# TABLE IX.

*Showing the Solubility of Salts.**

| Bases. / Acids. | Sulphuric. | Phosphoric. | Boracic. | Carbonic. | Silicic. | Arsenious. | Arsenic. | Hydrochloric. | Hydriodic. | Hydrosulphuric. | Nitric. | Chloric. | Oxalic. | Tartaric. | Citric. | Malic. | Succinic. | Benzoic. | Acetic. | Formic. |
|---|---|---|---|---|---|---|---|---|---|---|---|---|---|---|---|---|---|---|---|---|
| Potash | 1 | 1 | 1 | 1 | 1 & 2 | 1 | 1 | 1 | 1 | 1 | 1 | 1 | 1 | 1 & 2 | 1 | 1 | 1 | 1 | 1 | 1 |
| Soda | 1 | 1 | 1 | 1 | 1 . . 2 | 1 | 1 | 1 | 1 | 1 | 1 | 1 | 1 | 1 | 1 | 1 | 1 | 1 | 1 | 1 |
| Ammonia | 1 | 1 | 1 | 1 | .... | 1 | 1 | 1 | 1 | 1 | 1 | 1 | 1 | 1 & 2 | 1 | 1 | 1 | 1 | 1 | 1 |
| Magnesia | 1 | 2 | 2 | 2 | 2 & 3 | 2 | 2 | 1 | 1 | .... | 1 | 1 | 2 | 1 . . 2 | 1 | 1 . . 2 | 1 | 1 | 1 | 1 |
| Lime | 1 3 | 2 | 2 | 2 | 2 & 3 | 2 | 2 | 1 | 1 | 1 . . 2 | 1 | 1 | 2 | 2 | 2 | 1 & 2 | 2 | 1 . . 2 | 1 | 1 |
| Baryta | 3 | 2 | 2 | 2 | 2 & 3 | 2 | 2 | 1 | 1 | 1 | 1 | 1 | 2 | 2 | 2 | 1 | 1 . . 2 | 1 | 1 | 1 |
| Strontia | 3 | 2 | 2 | 2 | 2 & 3 | 2 | 2 | 1 | 1 | 1 | 1 | 1 | 2 | 2 | 1 | 1 | 1 | 1 | 1 | 1 |
| Alumina | 1 | 2 | 2 | .. | 2 & 3 | 2 | 2 | 1 | .. | .... | 1 | 1 | 2 | 1 | 1 & 2 | 2 | .... | 1 | 1 | 1 . . 2 |
| Ox. Chromium | 1 | 2 | 2 | 2 | 2 & 3 | .. | 1 | 1 | .. | .... | 1 | 1 | 1 | 1 | 1 | .... | 1 | .... | 1 | |
| Ox. Zinc | 1 | 2 | 2 | 2 | 2 & 3 | .. | 1 & 2 | 1 | 1 | 2 | 1 | 1 | 2 | 2 | 2 | 1 . . 2 | 1 | 1 | 1 | 1 |
| Ox. Manganese | 1 | 2 | 2 | 2 | 2 & 3 | .. | 2 | 1 | 1 | 2 | 1 | 1 | 2 | 1 . . 2 | 1 | 1 | 1 | 1 | 1 | 1 |
| Protox. Iron | 1 | 2 | 2 | 2 | 2 & 3 | 2 | 2 | 1 | 1 | 2 | 1 | 1 | 1 . . 2 | 1 . . 2 | 1 | 1 | 1 | .... | 1 | 1 |
| Perox. Iron | 1 | 2 | 2 | .. | 2 & 3 | .. | 2 | 1 | 1 | 2 | 1 | 1 | 1 . . 2 | 1 | 1 | 1 | 2 | 2 | 1 | 1 |
| Ox. Nickel | 1 | 2 | 2 | 2 | 2 & 3 | 2 | 2 | 1 | .. | 2 | 1 | 1 | 2 | 2 | 2 | .... | 2 | 1 | 1 | 1 |
| Ox. Cobalt | 1 | 2 | 2 | 2 | 2 & 3 | 2 | 2 | 1 | .. | 2 | 1 | 1 | 2 | 1 | .... | .... | 2 | 1 | 1 | 1 . . 2 |
| Ox. Antimony | 2 | .. | .. | .. | .... | 2 | 2 | 1 & 2 | .. | 2 | .... | .... | 1 . . 2 | 1 | .... | .... | .... | 1 | 1 | |
| Protox. Mercury | 1 2 | 2 | 1 | 2 | .... | 2 | 2 | 2 . . 3 | 2 | 3 | 1 & 2 | 1 & 2 | 2 | 1 . . 2 | 2 | 1 | 2 | 2 | 1 . . 2 | 1 |
| Perox. Mercury | 1 | 2 | .. | 2 | .... | 2 | 2 | 1 | 2 | 3 | 1 | 1 | 2 | 2 | .... | 1 & 2 | 1 . . 2 | 2 | 1 | |
| Ox. Lead | 3 | 2 | 2 | 2 | 3 | 2 | 2 | 1 . . 3 | 2 | 2 | 1 | 1 | 2 | 2 | 2 | 2 | 1 . . 2 | 1 | 1 | 1 |
| Ox. Copper | 1 | 2 | 2 | 2 | 3 | 2 | 2 | 1 | 2 | 2 | 1 | 1 | 2 | 1 | 1 | 1 | 1 | 1 . . 2 | 1 | 1 |
| Ox. Silver | 1 3 | 2 | 2 | 2 | .... | 2 | 2 | 3 | 3 | 2 | 1 | 1 | 2 | 2 | 2 | 1 | 1 | 1 | 1 | 1 |
| Protox. Tin | 1 | 2 | 2 | .. | .... | .. | 2 | 1 | 2 | 2 | .... | .... | 2 | 1 . . 2 | .... | 1 | 1 | 1 . . 2 | 1 | 2 |
| Perox. Tin | 1 | .. | .. | .. | .... | .. | .... | 1 | .. | 2 | 1 | .... | .... | .... | .... | .... | 2 | .... | 1 | |
| Ox. Bismuth | 1 | 2 | 2 | 2 | 3 . . | .. | 2 | 1 | 2 | 2 | 1 & 2 | .... | 2 | 2 | .... | .... | 1 | 1 | 1 | 1 |

* To ascertain the solubility of any salt, find the name of the base in the upright column, and that of the acid in the line at the top; the number placed at the point where the two rows meet, shows whether the salt formed by their combination is soluble or otherwise. The figure 1 means that it is soluble in water; 2, that it is insoluble in water, but soluble in acids; and 3, that it is insoluble in water and acids.

## TABLE X.

*Showing the Action of Reagents on Oxides and Acids. (Alphabetically arranged.)*

### 1. Metallic Bases. (In Combination.)

| *Name of Base.* | *Symbol.* | *Hydro-sulphuric Acid (Sulphuretted Hydrogen) (HS), in an acidified solution.* | *Hydro-sulphate of Ammonia ($NH_4S$, $HS$).* | *Carbonate of Soda. ($NaO$, $CO_2$).* | *Carbonate of Ammonia ($2NH_4O$, $3CO_2$).* | *Potash ($KO$).* | *Ammonia ($NH_3$).* | *Ferro-cyanide of Potassium ($K_2$, $Fe$ $Cy_3$).* | *Blowpipe.* | *Remarks.* |
|---|---|---|---|---|---|---|---|---|---|---|
| Alumina (see also p. 86). | $Al_2O_3$. | O | White; insol. in excess. | White; insol. in excess. | White; insol. | White; sol. in excess. | White; insol. | O | Blue with nitrate of cobalt. | These precipitates are insoluble in muriate of ammonia. |
| Antimony oxide of (see also) p. 106). | $SbO_3$. | Orange red. | Orange red; sol. in excess. | White; sparingly soluble. | White; sp. sol. | White; sp. sol. in. excess. | White; insol. | White. | With soda in deoxodizing flame, reduced, and gives off white fumes of oxide. | The chloride decomposed by water. |
| Baryta (see also p. 82). | $BaO$. | O | O | White; insol. in excess. | White; insol. | O | O | O | O | Thrown down immediately with sulphates and sulphate of lime. |
| Bismuth, oxide of (see also p. 118). | $Bi_2O_3$. | Brown-black. | Brown-black; insol. in excess. | White; insol. in excess. | White; insol. | White; insol. | White; insol. | White. | With soda on charcoal reduced; brittle bead of metal. | Nitrate decomposed by water. |

| | | | | | | | | | | |
|---|---|---|---|---|---|---|---|---|---|---|
| Cadmium, oxide of. | $CdO$. | Bright yellow. | Bright yellow; insol. in excess. | White; insol. in excess. | White; insol. | White; insol. | White; sol. in excess. | White. | With soda on charcoal reduced and metal volatilized; leaves a reddish brown deposit. | The yellow sulphide, insol. in hydrosulp. of ammonia, is highly characteristic. |
| Calcium (Lime) (see also p. 80). | $CaO$. | O | O | White; insol. in excess. | White; insol. | O | O | O | Radiates a brilliant light. Gives a red color to the flame. | Oxalate of ammonia causes a white precipitate even in very dilute solutions. |
| Chromium, oxide of (see also p. 87). | $Cr_2O_3$. | O | Green. | Green. | Green. | Green: sol. in excess. | Green; insol. | O | Emerald green with fluxes. | The oxide when fused with nitre, gives chromate of potash. |
| Cobalt, oxide of (see also p. 95). | $CoO$. | O | Black. | Pink; insol. Bluish on boiling. | Pink; sp. sol. in ex. solution purple. | Blue; becoming greenish; dirty red on boiling. | Blue; sol. in ex. forming a brownish red sol. | Pale green or gray. | Blue glass, with borax in both flames. | Readily distinguished with the blowpipe. |
| Copper, oxide of (see also p. 113). | $CuO$. | Black. | Black. | Greenish blue, becoming dark brown on boiling. | Greenish blue, sol. in ex. forming deep blue solution. | Pale blue, becoming dark brown when boiled. | Pale blue, sol. in ex. forming rich blue solution. | Mahogany colored; insol. | With soda on charcoal reduced. With borax and mic. salt in outer flame, green; in inner flame red. | Precipitated in the metallic state by clean iron; and as the black oxide by zinc. |
| Glucina. | $GlO_6$. | O | O | White; sp. sol. | White; sol. in ex. | White; sol. in ex. | White; insol. | O | With nitrate of cobalt, dark gray or black. | Glucina dissolves in cold solution of carb. ammonia, and is thrown down on boiling. |
| Gold, teroxide of. | $AuO_3$. | Black. | Brown-black; sol. in excess. | O | Yellow; insol. | Yellowish brown. | Yellow; insol. | O | Reduced. | Thrown down in the form of a brown metallic powder when boiled with protosulph. iron. |

| *Name of Base.* | *Sy mbo* | *Hydro-sulphuric Acid* (HS), *in an acidified solution.* | *Hydro-sulphate of Ammonia* ($NH_4S$, $HS$). | *Carbonate of Soda* ($NaO$, $CO_2$). | *Carbonate of Ammonia* ($2NH_4O$, $3CO_2$). | *Potash* ($KO$). | *Ammonia* ($NH_3$). | *Ferro-cyanide of Potassium* $K_2Fe$ ($Cy_3$). | *Blowpipe.* | *Remarks.* |
|---|---|---|---|---|---|---|---|---|---|---|
| IRIDIUM, sesquioxide of. | $Ir_2O_3$. | Slight brown. | Brown; sol. | Brown-red; sp. sol. | Bleaches the solution. | Slight brown. The solution becomes first colorless, and subsequently bluish. | As with potash. On exposure to the air, a slight blue precipitate falls. | Solution slowly discolored. | Reduced. | The solutions have a deep brown color. |
| IRON, protoxide of (see also p. 91). | $FeO$. | O | Black. | White, then green, and ultimately rust-colored. | As with carbonate of soda. | White becoming green. and on standing, rust-colored. | As with potash; but becoming brown more rapidly. | White; instantly changing to light blue. | With the fluxes in the outer flame, brownish-yellow; in the inner flame, light green. | Deep blue precipitate with ferridcyanide of potassium. |
| IRON, peroxide of (see also p. 92). | $Fe_2O_3$. | Yellowish-white precipitate of sulphur. | Black. | Rust-colored. | Rust-colored. | Rust-colored. | Rust-colored. | Deep-blue. | As the protoxide. | Black with infusion of gall-nuts. |
| LEAD, oxide of (see also p. 111). | $PbO$. | Black. | Black. | White; insol. | White; insol. | White; sol. | White; insol. None at first with the acetate. | White. | With soda on charcoal, reduced; yellow deposit also formed on the charcoal. | Preciptd. by soluble sulphates, and the precip. blackened by hydrosulphate of ammonia. Bright yellow with chromate of potash and iodide of potassium. |
| LITHIA. | $LiO$. | O | O | Faint white in concentrated sol. | As carb. soda. | O | O | O | Gives red color to the flame. | Phos. of soda and ammonia gives a white precipitate. |

| | | | | | | | | | | |
|---|---|---|---|---|---|---|---|---|---|---|
| MAGNESIA (see also p. 78). | $MgO$. | O | O | White insol. | O | White; insol. | White; insol. | O | Light pink with nitrate of cobalt. | Crystalline precipit. with phosphate of soda and ammonia. The carbonate and hydrate sol. in muriate of ammonia. |
| MANGANESE, protoxide of (see also p. 89). | $MnO$ | O | Flesh colored. | White; insol. | White; insol. | White, becoming brown. | White; becoming brown. | White. | With soda, a green bead. With borax in outer flame an amethyst bead, which loses its color in the reducing flame. | The presence of ammoniacal salts prevents more or less completely the precipitation of manganese by the alkalies. |
| MERCURY, protoxide of (see also p. 107). | $HgO$. | Black. | Black. | Dark-gray. | Dark-gray. | Black; insol. | Black; insol. | White. | Mixed with soda and heated in a tube, the metal sublimes. | White precipitate with chlorides, blackened by ammonia. Volatd. or decompd. by heat. |
| MERCURY, peroxide of (see also p. 109). | $HgO_2$. | White, turning to black. | White, turning to black. | Reddish-brown; insol. | White insol. | Yellow; insol. | White; insol. | White. | As the protoxide. | Volatilized or decompd. by heat. Beautiful scarlet with iodide of potassium. |
| MOLYBDENUM, oxides of. | $MoO$ and $MoO_2$ | Brown-black, slowly formed. | Yellowish-brown; sol. | Brown; sol. | Brown; sol. | Brown-black; insol. | Brown-black; insol. | Brown-with the binoxide. | With microcosmic salt in outer flame, a green glass. | Most readily distinguished by the blowpipe. |
| NICKEL, oxide of (see also p. 94). | $NiO$. | O | Black. | Pale green; insol. | Pale green; soluble, forming green solution. | Pale green; insol. | Pale green; soluble, forming a blue solution. | Pale green. | With soda on charcoal, reduced to a magnetic powder. With borax and mic. salt in outer flame, red glass, becoming colorless on cooling. | Potash throws down a pale-green precipitate from the ammoniacal solution. |

| Name of Base. | Symbol. | Hydro-sulphuric Acid (HS), in an acidified solution. | Hydro-sulphate of Ammonia ($NH_4S,HS$). | Carbonate of Soda ($NaO, CO_2$). | Carbonate of Ammonia ($NH_4O$, $3CO_2$). | Potash ($KO$). | Ammonia ($NH_3$). | Ferrocyanide of Potassium ($K_2$, $Fe Cy_3$). | Blowpipe. | Remarks. |
|---|---|---|---|---|---|---|---|---|---|---|
| OSMIUM, deutoxide of. | $OsO_2$. | Yellowish-brown, slowly formed. | Yellowish-brown insol. | Black, slowly formed. Bluish solution. | Brown after some time. | Black on boiling. | Brown after some time. | O | Osmium is characterized by forming, when heated in the air, a suboxide, which is volatile, and has a very disagreeable smell, causing much inconvenience to the eyes and nose. | |
| PALLADIUM, protoxide of. | $PdO$. | Black. | Black; insol. | Brown; sol. Reprecipitated on boiling. | Solution decolorized, but no precipitate. | Yellowish-brown; sol. | Yellowish brown sol. | O | Reduced. | Yellowish-white with solution of cyanide of mercury. |
| PLATINUM, oxide of. | $PtO_2$. | Brownish-black formed slowly. | Brownish-black; sol. in large excess. | Yellow with carbonate of potash. | Yellow. | Yellow. | Yellow. | O | Reduced. | Yellow with muriate of ammonia, which is converted by heat into spongy platinum. |
| POTASH, (see also p. 73). | $KO$. | O | O | O | O | O | O | O | Violet flame. | White crystalline precipitate with tartaric acid. Yellow with bichloride of platinum. |
| RHODIUM, sesquioxide of. | $R_2O_3$. | Brown, formed slowly. | Brown insol. | Yellowish after a time. | Yellowish after a time. | Yellowish brown on boiling. | Yellowish after a time. | Dark orange. | Reduced. | Many of the compounds have a rose-color. |
| SILVER, oxide of (see also p. 114). | $AgO$. | Black. | Black. | White; insol. | White; sol. | Pale brown; insol. | Pale brown sol. | White. | Reduced. | White curdy precipitate, with hydrochloric acid and chlorides, which is sol. in ammonia, and insol. in nitric acid. |

| | | | | | | | | | | |
|---|---|---|---|---|---|---|---|---|---|---|
| SODA (see also p. 74). | $NaO$. | O | O | O | O | O | O | O | Yellow flame. | The only salt which precipitates soda, is the antimoniate of potash. Evaporated with bichloride of platinum, gives yellow needles. |
| STRONTIA (see also p. 84). | $SrO$. | O | O | White; insol. | White; insol. | O | O | O | Carmine flame. | White precipitates with sulphates. Burnt with alcohol, gives carmine flame. |
| TIN, protoxide of (see also p. 115). | $SnO$. | Brown black. | Brown black. | White; insol. | White; insol. | White; sol. | White; insol. | White. | With soda in reducing flame, a malleable bead of metallic tin. | Zinc throws down the metal in beautiful crystals. |
| TIN, peroxide of (see also p. 116). | $SnO_2$. | Yellow. | Yellow sol. | White; insol. | White; insol. | White; sol. | White; sol. | White. | Reduced with soda. | The behavior with hydrosulphate of ammonia and the blowpipe are characteristic. |
| URANIUM, sesquioxide of. | $U_2O_3$. | O (Sulphur.) | Black. | Yellow; sol. | Yellow; sol. | Yellow; insol. | Yellow; insol. | Reddish-brown. | Yellow glass with borax. | When the precipitate with ammonia is heated, it is converted into the green protoxide. |
| VANADIUM, binoxide of. | $VO_2$. | O | Brown-black; sol. in excess, forming a purple solution. | Dirty white | Gray, passing to brown. | Grayish-white. | Brown. | Yellow. | With borax, yellow in outer flame; in the inner, brown, becoming green when cold. | Many of the solutions have a blue color. |

| *Name of Base.* | *Symbol.* | *Hydrosulphuric Acid* (HS), *in an acidified solution.* | *Hydrosulphate of Ammonia* ($NH_4S$, $HS$). | *Carbonate of Soda* ($NaO$, $CO_2$). | *Carbonate of Ammonia* ($2NH_4O$, $3CO_2$). | *Potash.* ($KO$). | *Ammonia* ($NH_3$). | *Ferrocyanide of Potassium* ($K_2.Fe\ Cy_3$). | *Blowpipe.* | *Remarks.* |
|---|---|---|---|---|---|---|---|---|---|---|
| YTTRIA. | $YO$. | O | White. | White; sp. sol. | White; sp. sol. | White; insol. | White; insol. | White. | Nothing characteristic. | Copious white with oxalic acid. |
| ZINC, oxide of (see also p. 88). | $ZnO$. | O | White. | White; insol. | White; sol. | White; sol. | White; sol. | White. | With soda on charcoal gives a white sublimate of oxide, which is yellow when hot. With nit. cobalt, green. | Behavior with hydrosulphate of ammonia characteristic. |
| ZIRCONIA. | $Zr_2O_3$ | O | White. | White after a time. | White; sol. | White; insol. | White; insol. | White. | Bright flame. | Oxalic acid gives a white precipitate. |

2. METALLIC OXIDES HAVING ACID PROPERTIES.

| *Acids (in combination).* | *Symbol.* | *Hydrosulphuric acid* (HS) *in acidified solutions.* | *Hydrosulphate of Ammonia* ($NH_4S$, $HS$). | *Chloride of Barium* ($BaCl$), *(in alkaline salts of the acids).* | *Nitrate of Silver* ($AgO$, $NO_5$), *(in alkaline salts of the acids).* | *Nitrate of Lime* ($CaO$, $NO_5$), *(in alkaline salts of the acids).* | *Hydrochloric acid* ($HCl$). | *Remarks.* |
|---|---|---|---|---|---|---|---|---|
| ANTIMONIOUS ACID. | $SbO_4$. | Orange. | Orange; sol. | White; sp. sol. in. water. | White. | White; sp. sol. in water. | White. | Antimonious acid becomes pale-yellow when heated, and white again on cooling. It is insoluble in nitric acid, and difficultly soluble in hot hydrochloric acid. |

| | | | | | | | | |
|---|---|---|---|---|---|---|---|---|
| ANTIMONIC ACID. | $SbO_5$. | Orange. | Orange ; sol. | White. | White. | White. | White. | Insoluble in water and nitric acid. Soluble in hydrochloric acid, from which it is precipitated on the addition of water. When strongly heated, gives off oxygen, and becomes antimonious acid. |
| ARSENIOUS ACID (see also p. 97). | $AsO_3$. | Yellow ; soluble in alkalies and alkaline sulphides. | Yellow ; sol. | White. | Pale yellow. | White. | O | Volatilizes at a low heat, and condenses in octohedral crystals. The best tests are Marsh's and Reinsch's. (See p. 89.) |
| ARSENIC ACID (see also p. 97). | $AsO_5$. | Yellow ; soluble in alkalies and alkaline sulphides. | Yellow ; sol. | White. | Chocolate-brown. | White. | O | Heated with black flux, gives metallic arsenic. |
| CHROMIC ACID. | $CrO_3$. | Reduced to oxide with precipitation of sulphur. | Green. | Yellow. | Reddish-brown. | Yellow in concentrated solutions. | Reduced to oxide, with evolution of chlorine. | Is decomposed by heat and by deoxiding agents, into oxide of chromium. Salts of lead throw down a yellow precipitate. |
| MANGANIC ACID. | $MnO_3$. | O | Flesh-colored. | O | Black (oxide). | Black. | Solution becomes red and chlorine is evolved. | Converted by acids into hypermanganic acid and peroxide of manganese ; the color of the solution changing from green to red. |
| MOLYBDIC ACID. | $MoO_3$. | Brown. | Brown : sol. | White. | White. | White. | White. | With microcosmic salt before the blowpipe, gives a dark-blue glass, which becomes green on cooling. When strongly heated, molybdic acid volatilizes and condenses in crystals. |
| TUNGSTIC ACID. | $WO_3$. | Slight turbidity. | Brown ; sol. | White. | White. | White. | White; insol. | Does not volatilize when heated. Has a pale yellow color, and is insoluble in water and acids. |

| Acids (in combination). | Symbol. | Hydrosulphuric acid (HS) in acidified solutions. | Hydrosulphate of Ammonia ($NH_4S$, $HS$). | Chloride of Barium ($BaCl$). (in alkaline salts of the acids). | Nitrate of Silver ($AgO$, $NO_5$), (in alkaline salts of the acids). | Nitrate of Lime ($CaO$, $NO_5$), (in alkaline salts of the acids. | Hydrochloric acid ($HCl$). | Remarks. |
|---|---|---|---|---|---|---|---|---|
| VANADIC ACID. | $VO_3$. | Gray. | Brown; sol. | Orange. | Yellow. | O | Chlorine. evolved. | When treated with hydrochloric acid, the mixture is capable of dissolving gold-leaf. Vanadic acid in solution is readily deoxidized, forming a blue liquid. |

3. NON-METALLIC ACIDS.

| Acids (neutralized). | Symbol. | Nitrate of Baryta ($BaO$, $NO_5$). | Nitrate of Silver ($AgO$, $NO_5$). | Nitrate of Lime ($CaO$, $NO_5$). | Acetate of Lead ($PbO$, $C_4H_3O_3$). | Remarks. |
|---|---|---|---|---|---|---|
| BORACIC ACID (see also p. 122). | $BO_3$. | White. | White. | White. | White. | Slightly volatile in the presence of aqueous vapor. Turns turmeric-paper brown, and blue litmus port-wine color. Gives green color to the flame of alcohol. |
| BROMIC ACID. | $BrO_5$. | White. | White. | O | White. | The bromates are decomposed by heat into bromides and oxygen. Sulphuric acid disengages bromine. |
| CARBONIC ACID (see also p. 123). | $CO_2$. | White. | White. | White. | White. | The carbonates are readily decomposed by acids, carbonic acid gas being given off with effervescence, which, when passed into lime-water, gives a white precipitate. |
| CHLORIC ACID (see also p. 130). | $ClO_5$. | O | O | O | O | All the chlorates are soluble in water. At a red heat they are converted into chlorides, oxygen being given off. |
| HYDRIODIC ACID (see also p. 126). | $HI$. | O | Pale yellow. | O | Bright Yellow. | The iodides evolve iodine when heated with nitric or sulphuric acid. With chlorine water and starch, they give a dark-purple precipitate. |

| | | | | | | |
|---|---|---|---|---|---|---|
| Hydrobromic acid. | $HBr$. | O | Yellowish. | O | White. | The bromides, when heated with nitric acid, evolve bromine. |
| Hydrochloric acid (see also p. 125). | $HCl$. | O | White. | O | White. | The chlorides, when heated with peroxide of lead, or of manganese, evolve chlorine. |
| Hydrocyanic acid. | $H,C_2N$. | O | White. | O | White. | With a mixture of protosalt, and persalt of iron, the alkaline cyanides give a precipitate of Prussian blue. |
| Hydrofluoric Acid. | $HF$. | White. | White. | White. | White. | The fluorides, when moistened with sulphuric acid, give off fumes which corrode glass. |
| Hydroselenic acid. | $HSe$. | O | Black. | O | Black. | The selenides, when heated in the outer flame of the blowpipe, evolve the odor of selenium, resembling that of putrid horse-radish. |
| Hydrosulphuric acid (see also p. 127). | $HS$. | O | Black. | O | Black. | Most of the sulphides, when treated with an acid, evolve hydrosulphuric acid, which smells like rotten eggs. |
| Hyposulphurous acid. | $S_2O_2$. | White. | White becoming brown. | O | White. | The hyposulphites are decomposed by hydrochloric acid; sulphur is precipitated, and sulphurous acid set free. |
| Hyposulphuric acid. | $S_2O_5$. | O | O | O | O | The hyposulphates are decomposed without deposition of sulphur, when boiled with hydrochloric acid; sulphurous and sulphuric acids are formed. |
| Iodic acid. | $IO_5$. | White. | White. | White. | White. | The iodates are decomposed by heat into iodides and oxygen. |
| Nitric Acid (see also p. 129). | $NO_5$. | O | O | O | O | When mixed with sulphuric and hydrochloric acids, the nitrates dissolve gold leaf. With copper filings and sulphuric acid, orange fumes are given off. |
| Perchloric acid. | $ClO_7$. | O | O | O | O | The perchlorates are resolved by heat into chlorides and oxygen. They are not decomposed in the cold by hydrochloric or sulphuric acid; thus differing from the chlorides. |
| Phosphoric acid (Tribasic) (see also p. 120). | $PO_5$. | White. | Pale yellow. | White. | White. | The soluble phosphates give with salts of magnesia, when ammonia is present, a white crystalline precipitate. |

| *Acids (neutralized).* | *Symbol.* | *Nitrate of Baryta ($BaO$, $NO_5$).* | *Nitrate of Silver ($AgO$, $NO_5$).* | *Nitrate of Lime ($CaO$, $NO_5$).* | *Acetate of Lead ($PbO$ $C_4H_3O_3$).* | *Remarks.* |
|---|---|---|---|---|---|---|
| PHOSPHOROUS ACID. | $PO_3$. | White. | White, becoming brown. | White. | White. | The hydrated phosphites are decomposed when heated in a tube; hydrogen is given off, and phosphates are formed. |
| SELENIC ACID. | $SeO_3$. | White. | White. | White. | White. | The seleniates are decomposed by boiling with hydrochloric acid; chlorine is evolved, together with selenious acid. |
| SELENIOUS ACID. | $SeO_2$. | White. | White. | White. | White. | Metallic zinc or sulphurous acid causes the precipitation of selenium from acidified solutions of the selenites. |
| SILICIC ACID (see also p. 124). | $SiO_3$. | White. | Pale yellow. | White. | White. | When a soluble silicate is evaporated to dryness with hydrochloric acid, it is decomposed, and the silica remains insoluble. |
| SULPHURIC ACID (see also p. 119). | $SO_3$. | White. | White crystalline. | White crystalline. | White. | Most of the sulphates, when heated with charcoal, are converted into sulphides, which, when moistened with hydrochloric acid, evolve hydrosulphuric acid. |
| SULPHUROUS ACID. | $SO_2$. | White. | White. | White. | White. | The sulphites are decomposed by sulphuric acid, sulphurous acid being given off, without the deposition of sulphur. |

4. ORGANIC ACIDS.

| *Acids (neutralized).* | *Symbol.* | *Chloride of Calcium ($CaCl$).* | *Perchloride of Iron ($Fe_2Cl_3$).* | *Nitrate of Baryta ($BaO$, $NO_5$).* | *Nitrate of Silver ($AgO,NO_5$).* | *Acetate of Lead ($PbO$, $C_4H_3O_3$).* | *Remarks.* |
|---|---|---|---|---|---|---|---|
| ACETIC ACID (see also p. 138). | $HO,C_4.H_3O$. | O | O | O | White crystalline in concentrated solutions. | O | The acetates, when warmed with sulphuric acid, give off the smell of vinegar. Acetic acid boiled with an excess of protoxide of lead, forms the subacetate, which is alkaline to test-paper. |

| | | | | | | | |
|---|---|---|---|---|---|---|---|
| Benzoic acid (see also p. 137). | $HO$, $C_{14}H_5O_3$. | O | Brownish yellow. | O | Crystalline in concentrated neutral solutions. | White in concentrated neutral solutions. | Solutions of the benzoates, when treated with sulphuric acid, give a crystalline precipitate of benzoic acid. |
| Citric acid (see also p. 135). | $3HO$, $C_{12}H_5O_{11}$. | White. | O | White. | White. | White. | With protonitrate of mercury, a white precipitate, which becomes gray. |
| Formic acid (see also p. 139). | $HO$, $C_2HO_3$. | O | O | O | White; becoming black, especially when warmed. | O | The formiates, when warmed with sulphuric acid, do not blacken, and give off carbonic oxide gas. |
| Malic acid (see also p. 135). | $2HO$. $C_8H_4O_8$. | White on the addition of alcohol. | O | White. | White; becoming gray. | White precipitate, that melts in boiling water. | Malate of lead dissolves in hot dilute acetic acid, and crystallizes on cooling in fine needles. Malic acid is decomposed by heat, into malæic and fumaric acids. |
| Oxalic acid (see also p. 122). | $HO$, $C_2O_3$. | White. | Yellowish brown. | White crystalline. | White. | White. | Neither the acid nor the oxalates are blackened by strong sulphuric acid, but give off carbonic acid and carbonic oxide gases. |
| Succinic acid (see also p. 136). | $HO$, $C_4H_2O_3$. | O | Reddish brown. | O | White on standing. | White. | A mixture of chloride of barium, ammonia, and alcohol, gives a white precipitate of succinate of baryta. |
| Tartaric acid (see also p. 133). | $2HO$, $C_8H_4O_{10}$. | White. | O | White. | White. | White. | Added in excess to potash, gives a crystalline precipitate of the bitartrate. |

## TABLE XI.

*Showing the Behavior of Solutions of the Metals with Hydrosulphuric Acid, Hydrosulphate of Ammonia, and Carbonate of Ammonia, employed successively. (Dr. Will.)—(The rarer metals are printed in italics.)*

| Elements precipitated from their acid solutions by Hydrosulphuric Acid, as Sulphides. | | | | Bodies precipitated by Hydrosulphate of Ammonia. | | | | | Bodies not precipitated by Hydrosulphuric Acid, or Hydrosulphate of Ammonia. In the presence of Muriate of Ammonia, on addition of Carbonate of Ammonia, are | |
|---|---|---|---|---|---|---|---|---|---|---|
| Soluble in Hydrosulphate of Ammonia, and reprecipitated by Hydrochloric Acid. | | Insoluble in Hydrosulphate of Ammonia. | | As Sulphides. | | As Oxides. | | As Salts. | precipitated. | not precipitated. |
| Antimony | Orange. | Mercury | Black or Brownish-black. | Nickel | Black. | Alumina | Soluble in potash. | Baryta, Strontia, Lime, in combination with phosphoric, boracic, oxalic, and some other acids. | Baryta. | Magnesia. |
| Arsenic | Yellow. | Silver | Black or Brownish-black. | Cobalt | Black. | *Glucina* | Soluble in potash. | | Strontia. | Potash. |
| Tin | Yellow. | Lead | Black or Brownish-black. | Manganese | Flesh-colored. | Chromium | Soluble in potash. | | Lime. | Soda. |
| *Gold* | Black. | Bismuth | Black or Brownish-black. | Iron | Black. | *Thorina* | Insoluble in potash. | | | *Lithia.* |
| *Platinum* | Black. | Copper | Black or Brownish-black. | Zinc | White. | *Yttria* | Insoluble in potash. | Magnesia in combination with phosphoric acid. | | Ammonia. |
| *Iridium* | Black. | *Cadmium* | Yellow. | *Uranium* | Brownish-black. | *Cerium* | Insoluble in potash. | | | |
| *Molybdenum* | Brown. | *Palladium* | Brownish-black. | | | *Zirconia* | Insoluble in potash. | | | |
| | | *Rhodium* | Brownish-black. | | | *Titanium* | Insoluble in potash. | | | |
| | | *Osmium* | Brownish-black. | | | *Tantalium* | Insoluble in potash. | | | |

# LIST OF SALTS, ETC.

WHICH MAY BE EXAMINED FOR PRACTICE IN QUALITATIVE ANALYSIS (Part III).

---

### (*a.*) *Simple Salts, &c., soluble in Water.*

Chloride of barium.
Sulphate of soda.
Muriate of ammonia.
Sulphate of magnesia.
Chloride of calcium.
Nitrate of strontia.
Sulphate of chromium.
Sulphate of zinc.
Sulphate of manganese.
Protosulphate of iron.
Perchloride of iron.
Sulphate of nickel.
Nitrate of cobalt.
Protonitrate of mercury.
Perchloride of mercury.
Acetate of lead.
Sulphate of copper.
Nitrate of silver.
Phosphate of soda.
Iodide of potassium.
Biborate of soda.
Nitrate of potash.
Chlorate of potash.
Carbonate of soda.
Arsenious acid.

### (*b.*) *Simple Salts, &c., insoluble in Water, but soluble in Acids.*

Carbonate of magnesia.
Phosphate of lime.
Carbonate of baryta.
Metallic zinc.
Carbonate of strontia.
Sulphide of iron.
Protoxide of lead.
Sulphide of antimony.
Black oxide of copper.
Carbonate of lime.
Oxide of bismuth.
Metallic tin.

### (*c.*) *Simple Salts, &c., insoluble in Water and Acids.*

Sulphate of baryta.
Sulphate of strontia.
Chloride of lead.
Sulphate of lead.
Silica.
Chloride of silver.
Silicate of lime.
Silicate of alumina.

(*d.*) *Mixed Salts, &c., soluble in Water.*

{ Chloride of barium,
Nitrate of potash.

{ Sulphate of zinc,
Chloride of sodium.

{ Muriate of ammonia,
Phosphate of soda.

{ Sulphate of copper,
Nitrate of cobalt.

{ Perchloride of iron,
Sulphate of magnesia.

Alum (double sulphate of alumina and potash).

{ Chloride of calcium,
Nitrate of potash,
Muriate of ammonia.

{ Nitrate of lead,
Nitrate of cobalt,
Nitrate of strontia.

{ Sulphate of nickel,
Chloride of sodium,
Sulphate of magnesia.

{ Chlorate of potash,
Biborate of soda,
Muriate of ammonia,
Arsenious acid.

{ Nitrate of soda,
Nitrate of lime,
Nitrate of baryta,
Nitrate of zinc,
Nitrate of lead,
Nitrate of copper.

{ Sulphate of potash,
Phosphate of soda,
Biborate of soda,
Carbonate of ammonia,
Chloride of potassium,
Iodide of potassium,
Nitrate of ammonia.

(*e.*) *Mixed Salts, &c., insoluble in Water, but soluble in Acids.*

{ Carbonate of magnesia,
Sulphide of iron.

{ Protoxide of lead,
Phosphate of lime.

{ Carbonate of lime,
Black oxide of copper.

Brass.

{ Oxide of bismuth,
Sulphide of iron,
Sulphide of antimony.

Magnesian limestone.
Iron pyrites.
Copper pyrites.
Argentiferous galena.
German silver.
Arsenical cobalt ore.
The solid matter contained in sea, well, or river water.
The portion of soils which is soluble in acids.

(*f.*) *Mixed Salts, &c., insoluble in Water and Acids.*

{ Chloride of silver,
Sulphate of baryta.

{ Silica,
Chloride of lead.

{ Sulphate of lead,
Silicate of alumina,
Chloride of silver.

The insoluble portion of soils.
Slate.
Siliceous minerals.

# GLOSSARY OF CHEMICAL TERMS.[1]

ABSORPTION, from *absorbeo*, to suck up; the act of imbibing a liquid.

ACETIC ACID, from *acetum*, vinegar; the acid contained in vinegar.

AERIFORM, from ἀήρ, the air, and *forma*, a form; having the form or properties of air.

AFFINITY, from *ad*, to, and *finis*, a boundary; relationship; the force which causes particles of dissimilar kinds of matter to combine together, so as to form new matter.

ALBUMEN,-INOUS, from *albumen*, the white of an egg; an important animal principle. The white of an egg consists chiefly of albumen and water, contained in a cellular tissue.

ALCOHOL, from an Arabic word; the intoxicating principle of spirituous liquors.

ALKALI, a soluble body, with a hot caustic taste, which possesses the power of destroying or neutralizing acidity. The term is derived from the Arabic article, *al*, and *kali*, the Arabic name of a plant, from the ashes of which one of the most important alkalies (potash) is obtained.

AMALGAM, from ἅμα, together, and γαμέω, to marry; a term signifying the union of any metal with mercury, which has the property of dissolving several of the metals.

AMORPHOUS, from ἀ, not, and μορφὴ, a form; not possessing any regular form.

ANALOGUE, that which is the counterpart of another.

ANALOGY,-ICAL, and -OUS, from ἀνὰ, among, and λόγος, a relation or proportion; a likeness or resemblance between things, with regard to their circumstances or effects.

ANALYSIS, from ἀνὰ, among, and λύω, to loosen; the separation of a substance into its component parts.

ANGLE, from *angulus*, a corner; the inclination of two straight lines to each other, which meet together, but are not in the same straight line.

ANHYDROUS, from ἀ, not, and ὕδωρ, water; containing no water.

ANTISEPTIC, from ἀντὶ, against, and σήπω, to putrefy; possessing the power of preventing or retarding putrefaction.

AQUA REGIA, *i. e.*, REGAL WATER, a mixture of nitric and hydrochloric acids; so called from its property of dissolving gold, which was held by the alchemists to be the king of the metals.

[1] Many of the definitions are taken almost verbatim from Daniell's "Chemical Philosophy."

AQUEO, from *aqua*, water; when prefixed to a word, denotes that water enters into the composition of the substance which it signifies, as *aqueo-sulphuric acid* (*HO,SO* ).

ATHERMANOUS, from ἀ, not, and θέρμος, heat; that through which heat will not pass, is said to be *athermanous*.

ATMOSPHERE, from ἀτμός, vapor, and σφαῖρα, a sphere; commonly used to denote the sphere of air which surrounds the globe.

ATOM,-IC, from ἀ, not, and τέμνω, to cut; a minute particle of matter, not susceptible of further division.

ATTRACTION,-IVE, from *ad*, to, and *traho*, to draw; the tendency which bodies have to approach or unite with each other.

AZOTE, from ἀ, not, and ζωή, life; another name for nitrogen; so called because it is incapable of supporting respiration.

BARIUM, from βαρύς, heavy; the metallic base of baryta.

BAROMETER, from βάρος, weight, and μέτρον, a measure; an instrument for measuring the varying pressure of the atmosphere.

BARYTA, a compound of oxygen and the metal *barium* (BaO), possessing alkaline properties.

BIBULOUS, from *bibo*, to drink; that which has the property of drinking in, or absorbing, moisture.

BORON, a dark, olive-colored elementary substance, obtained from boracic acid, insoluble in water, and a non-conductor of electricity.

BROMINE, from βρῶμος, a strong odor; an elementary liquid of a reddish-brown color and suffocating smell; in chemical properties, it strongly resembles iodine.

CALORIC, from *calor*, heat; an imaginary fluid substance, supposed to be diffused through all kinds of matter, and the sensible effect of which is called *heat*.

CAPILLARY, from *capillus*, a hair, resembling, or having the form of hairs.

CAPSULE, from *capsula*, a little chest; a small shallow cup.

CARBON, from *carbo*, a coal; the chemical name for charcoal.

CAUSTIC, from καίω, to burn; possessing the power of burning.

CHEMISTRY,-ICAL, from an Arabic word, signifying the knowledge of the substance or constitution of bodies; the science whose object is to examine the constitution of bodies.

CHLORINE, from χλωρός, green; a greenish-colored gas, of a pungent suffocating smell, and possessing chemical properties nearly allied to those of oxygen.

CLEAVAGE, PLANE OF, the plane in which crystals have a tendency to separate.

COHESION, from *con*, together, and *hæreo*, to stick; the power which causes the particles of a body to cling together and resist separation.

COMBUSTION, from *comburo*, to burn; the disengagement of light and heat, which frequently accompanies chemical combination.

CONDUCTION, from *con*, together, and *duco*, to lead; the power of transmitting heat or electricity, without change in the relative position of the particles of the conducting body.

CONGELATION, from *con*, together, and *gelo*, to freeze; the process of freezing.

CONSTITUENT, from *constituo*, to put together; that of which anything consists, or is made up.

CONTRACTION, from *con*, together, and *traho*, to draw; the state of being drawn into a narrow compass, or becoming smaller.

CONVEX, from *con*, together, and *veho*, to carry; curved outwardly, or protuberant.

CORPUSCULAR, from *corpus*, a body; composed of, or relating to, atoms.

CRYSTALLOGRAPHY, from κρύσταλλος, a crystal or ice, and γράφω, to describe; the science which treats of crystals.

CRYSTALLIZATION; the formation of crystals during the passage of certain substances from a fluid to a solid state.

CUBE,-IC; a solid figure contained by six equal squares.

CYANOGEN, from κύανος, blue, and γεννάω, to produce; a colorless gas composed of carbon and nitrogen ($C_2N$). Its chemical properties much resemble those of oxygen and chlorine; it derives its name from the circumstance of its entering into the composition of Prussian blue.

CYANIDE; a compound of cyanogen with a metal is called cyanide, as cyanide of potassium ($K,C_2N$).

DECOMPOSITION; the resolution of a compound substance into its component parts.

DECREPITATION, from *de*, from, and *crepito*, to crackle; the crackling noise which certain salts make when heated, usually caused by the sudden escape of water.

DEFLAGRATION, from *deflagro*, to burn; burning.

DELIQUESCENCE, from *deliqueo*, to melt; a gradual melting or dissolving, caused by the absorption of water from the atmosphere.

DENSITY, from *densus*, thick; vicinity or closeness of particles; specific weight.

DEOXIDIZE; to deprive of oxygen.

DETONATION, from *detono*, to thunder; explosion accompanied with noise.

DIAPHANOUS, from διὰ, through, and φαίνω, to shine; that which allows a passage through to the rays of light, but disperses them so as to prevent direct vision.

DIATHERMANOUS, from διά, through, and θερμὸς, heat; that through which heat will pass, is said to be *diathermanous*.

DIMORPHOUS, from δὶς, twice, and μορφὴ, a form; having two distinct crystalline forms.

DISINTEGRATION, from *dis*, meaning separation, and *integer*, whole; a complete separation of particles.

DISTILLATION, a separation drop by drop; the process by which a fluid is separated from another substance by first being converted into vapor, and afterwards condensed drop by drop.

DODECAHEDRON, from δώδεκα, twelve, and ἕδρα, a base or side; a solid figure contained by twelve equal sides.

EBULLITION, from *ebullio*, to boil; the act of boiling.

Effervescence; the escape of bubbles of gas formed in a liquid, as when marble is decomposed by hydrochloric acid.

Efflorescence, from *effloresco*, to blow as a flower; the formation of small crystals on the surfaces of bodies, in consequence of the abstraction of water from them by the atmosphere.

Electricity, from ἤλεκτρον, amber; the name of a power of matter, which produces a variety of peculiar phenomena, the first of which were observed in the mineral substance called amber; the laws, hypotheses, and experiments by which they are explained and illustrated constitute the science of electricity.

Electrode, from ἤλεκτρον, electricity, and ὁδὸς, a way; the point at which an electric current enters or quits the body through which it passes.

Electrolysis, -lyte, from ἤλεκτρον, electricity, and λύω, to loosen; the act of decomposing bodies by electricity.

Element, from *elementum*, an element; that which cannot be resolved into two or more parts, and contains but one kind of ponderable matter.

Empyreumatic, from ἐν, in, and πῦρ, fire; having the taste or smell of burnt animal or vegetable substances.

Endosmose, from ἔνδον, within, and ὠσμὸς, the act of pushing; a flowing from the outside to the inside.

Equivalent, from *æquus*, equal, and *valeo*, to be worth; equal in value, or in the power of combining with other substances.

Evaporation, from *e*, out, and *vapor*, vapor; the conversion of a liquid into vapor.

Exosmose, from ἔξω, without, and ὠσμὸς, the act of pushing; a flowing from the inside to the outside.

Expansion, from *expando*, to open out; the enlargement or increase in the bulk of bodies, which is produced by heat.

Experiment, from *experior*, to attempt, to try; something done in order to discover an uncertain or unknown effect.

Explosion, from *ex*, out, and *plaudo*, to utter a sound; a sudden expansion of an elastic fluid, with force and a loud report.

Fermentation, from *fermentum*, that which is light and puffy; originally applied to the process by which alcohol is formed in saccharine liquids.

Ferruginous, from *ferrum*, iron; belonging to, or resembling, iron.

Filter, a strainer.

Fluorine, from *fluo*, to flow; an elementary principle contained in fluor spar, which is so called from its acting as a flux in the working of certain minerals.

Flux, from *fluo*, to flow; that which itself readily melts when heated, and assists in the fusion of other substances when mixed with it.

Focus, from *focus*, a fire-place; a point in which a number of rays of light or heat meet, after being refracted or reflected.

Gas, a permanent aeriform fluid.

Gelatinous, from *gelo*, to freeze; resembling jelly.

Gravity, from *gravis*, heavy; the natural tendency of bodies to fall towards a centre, usually the centre of the earth.

Gravity, Specific; the relative gravity or weight of a body, considered with regard to an equal bulk of some other body, which is assumed as a standard of comparison.

Heterogeneous, from ἕτερος, different, and γένος, kind; different in nature and properties.

Homogeneous, from ὁμός, like, and γένος, kind; alike in nature and properties.

Hydrate, from ὕδωρ, water; any substance which contains water chemically combined.

Hydrogen, from ὕδωρ, water; and γεννάω, to produce; an inflammable, colorless, and aeriform fluid; the lightest of all known substances, and one of the elements of water.

Hydro; when prefixed to the name of a chemical substance, denotes that hydrogen enters into the composition of the substance which it signifies.

Hydrostatics, from ὕδωρ, water, and στατός, poised; the branch of Natural Philosophy which treats of the pressure and equilibrium of non-elastic fluids, and also of the weight, pressure, &c., of solids immersed in them.

Hypo, from ὑπό, under; when prefixed to a word, denotes an inferior quantity of some ingredient which enters into the composition of the substance which it signifies.

Hypothesis, from ὑπό, under, and τίθημι, to place; a principle supposed, or taken for granted, in order to prove a point in question.

Ignite, from *ignis*, fire; to heat a substance to redness; to set on fire.

Imponderable, from *in*, not, and *pondero*, to weigh; that which has no perceptible weight.

Incandescent, from *incandesco*, to grow white; white or glowing with heat.

Increment, from *incresco*, to increase; the quantity by which anything increases or becomes greater.

Induction, electrical, from *in*, to, and *duco*, to lead; the effect produced by the tendency of an insulated electrified body, to excite an opposite electric state in neighboring bodies.

Inertia, from *inertia*, inactivity; the disposition of matter to remain in its state of rest or motion.

Inflammable, from *in*, and *flamma*, a flame; capable of burning with a flame.

Insulation, from *insula*, an island; when a body containing a quantity of free heat or electricity is surrounded by non-conductors, it is said to be *insulated*.

Interstices, from *interstitium*, a break or interval; the unoccupied spaces between the molecules of bodies.

Iodide; a compound of iodine and a metal.

Iodine, from ἴον, a violet, and εἶδος, the form or likeness; a soft opaque elementary substance, which, when heated, sublimes in the form of a violet-colored vapor.

Isomeric, from ἴσος, equal, and μέρος, a part; substances which consist of the same ingredients, in the same proportions, and yet differ essentially in their properties, are called *isomeric*.

Laminæ, from *lamina*, a thin plate; extremely thin plates, of which some solid bodies are composed.

Levigation, from *lævis*, smooth; the reducing of hard bodies to a very fine powder, by grinding with water.

Lignin, from *lignum*, wood; an organic principle of which the fibres of vegetables are mainly composed.

Litmus, a blue pigment obtained from the lichen *roccella;* it is a most delicate test for acids, which turn it red.

Malleable, from *malleus*, a hammer; that which is capable of being spread out by hammering.

Metallurgy, from μέταλλον, a metal, and ἔργον, a work; the art of working metals, and separating them from their ores.

Molecules, -ar, a diminutive from *moles*, a mass; the infinitely small material particles, of which bodies are conceived to be aggregations.

Mucilaginous; resembling mucilage or gum.

Murexide, from *murex*, a fish affording a purple dye; a beautiful purple compound, resulting from the decomposition of uric acid by means of nitric acid.

Nascent, from *nascor*, to be born; in the moment of formation.

Nitrogen, from νίτρον, nitre, and γεννάω, to produce; a colorless elementary gas, devoid of taste and smell; it is one of the constituents of the atmosphere, and also of nitric acid, from which latter circumstance it derives its name.

Nitrogenous; containing nitrogen in combination.

Nucleus, from *nucleus*, a kernel; the central parts of a body, which are supposed to be firmer, and separated from the other parts, as the kernel of a nut is from the shell; also, the point about which matter is collected.

Octohedron, -al, from ὀκτὼ, eight, and ἕδρα, a side; a solid figure contained by eight equal and equilateral triangles.

Olefiant gas, from *oleum*, oil, and *fio*, to become; a colorless gas, composed of carbon and hydrogen ($C_4H_4$), which derives its name from its property of forming an oil-like liquid with chlorine.

Organic matter, from ὄργανον, an organ; matter of which the organic parts or juices of plants and animals are composed, or which is derived from such parts by the action of chemical agents, is called *organic*.

Oxide; a compound of oxygen with a metal or non-metallic body, not having acid properties.

Oxidize; to combine with oxygen.

Oxygen, from ὀξὺς, acid, and γεννάω, to produce; a colorless elementary gas, which was formerly supposed to be the universal acidifying principle.

PELLICLE, a diminutive from *pellis*, a skin or crust; a thin crust formed on the surface of a solution by evaporation.

PERCOLATE, from *per*, through, and *colo*, to strain; to strain through.

PERMEATE, from *permeo*, to pass through; to penetrate.

PHENOMENON, from φαίνομαι, to appear; an appearance which is more or less remarkable.

PHILOSOPHY; from φιλέω, to love, and σοφία, wisdom; the study or knowledge of nature or morality, founded on reason and experience; the word originally implying "A love of wisdom."

PHLOGISTON, from φλέγω, to burn; a name given by the older chemists to an imaginary substance, which was considered as the principle of inflammability.

PHOSPHORUS, from φῶς, light, and φέρω, to produce; a highly inflammable elementary substance, obtained from calcined bones, which emits light when placed in the dark, owing to its undergoing a slow combustion.

PHYSICS, -ICAL, from φύσις, nature; the science of natural bodies, their phenomena, causes, and effects, with their affections, motions, and operations.

PNEUMATICS, from πνεῦμα, air; that branch of Natural Philosophy which treats of the weight, elasticity, and other properties of aeriform fluids.

POLARITY; the opposition of two equal forces in bodies, similar to that which confers the tendency of magnetized bodies to point towards the magnetic poles.

POLARIZED LIGHT; light, which by reflection or refraction at a certain angle, or by refraction in certain crystals, has acquired the property of exhibiting opposite effects in planes at right angles to each other, is said to be *polarized*.

PORES, from πόρος, a passage; the small interstices between the solid particles of bodies.

PRECIPITATION, from *præcipito*, to fall suddenly; the formation and separation of a solid substance in a liquid.

PRODUCT, from *pro*, forth, and *duco*, to draw; anything formed from the elements of another by an operation.

PYRO, from πῦρ, fire; when prefixed to a word, denotes that the substance which it signifies, has been formed at a high temperature.

QUALITATIVE; regarding the properties of a body, and the kinds of matter of which it is composed without reference to quantity.

QUANTITATIVE; regarding quantities.

RADIATION, from *radius*, a ray; the shooting forth in all directions from a centre.

RAREFACTION, from *rarus*, rare, and *facio*, to make; the act of causing a substance to become less dense; it also denominates the state of this lessened density.

RECTIFICATION; the process of drawing anything off by distillation, in order to obtain it in a state of greater purity.

REFRACTION, from *re*, back, and *frango*, to break, the deviation of rays of light or heat from their direct course, when passing through media of different densities.

Refrigeration, from *re*, again, and *frigus*, cold; the act of cooling.
Repulsion, from *re*, back, and *pello*, to drive; that property in certain bodies, whereby they mutually tend to recede from each other.

Salifiable bases, from *sal*, salt, and *fio*, to become; bodies capable of combining with acids, to form salts.
Sapid, from *sapio*, to taste of; possessing the power of exciting the organs of taste.
Saturation, -ated, from *satur*, full; the solution of one body in another until the receiving body can contain no more. A solution is said to be saturated with an acid or an alkali, when the latter is added in sufficient quantity to render it neutral, and *supersaturated* when the point of neutrality has been exceeded.
Solution, from *solvo*, to loosen or melt; any liquid which contains another substance dissolved in it.
Solvent, any substance which will dissolve another.
Specific, from *species*, a particular sort or kind; that which denominates any property which is not general, but is confined to an individual or species.
Specific gravity; see Gravity, specific.
Sublimation, from *sublimis*, high; the act of raising into vapor by means of heat, and condensing in the upper part of the vessel.
Sulphide; a combination of a metal with sulphur.
Supersaturate; see Saturation.

Ternary, from *ter*, thrice; containing three units.
Tetrahedron, from *τέσσαρες*, four, and *ἕδρα*, a base or side; a solid figure contained by four equal and equilateral triangles.
Transparent; a term to denote the quality of a substance which not only admits the passage of light, but also of the vision of external objects.
Triturate, from *trituro*, to thresh; to reduce to powder.

Vacuum, from *vacuus*, empty; a space empty, and devoid of all matter.
Volume, from *volumen*, a roll; the apparent space occupied by a body.

Weight; the pressure which a body exerts vertically downwards, in consequence of the action of gravity.

Zero; the numeral 0, which fills the blank between the ascending and descending numbers of a series.

# INDEX.

## A.

## D.

E.



F.



G.

## H.



## I.

## N.



## O.

## P.



## Q.

## R.



## S.

## T.

THE END.

# CATALOGUE

OF

# MEDICAL, SURGICAL, AND SCIENTIFIC WORKS

PUBLISHED BY

# BLANCHARD & LEA, PHILADELPHIA.

---

AMERICAN JOURNAL OF THE MEDICAL SCIENCES.—Edited by ISAAC HAYS, M. D. Published Quarterly, each number containing about 300 large octavo pages. Price, $5 per annum. When paid for in advance, it is sent free by post, and the "Medical News and Library," a monthly of 32 large 8vo. pages, is furnished gratis. Price of the "Medical News," separate, $1 per annum, in advance.

---

ALLEN (J. M.), M. D.—THE PRACTICAL ANATOMIST; OR, THE STUDENT'S GUIDE IN THE DISSECTING-ROOM. With over 200 illustrations. In one handsome royal 12mo. volume. (*Just Issued.*)

---

ABEL (F. A.), F. C. S., and C. L. BLOXAM.—HANDBOOK OF CHEMISTRY, Theoretical, Practical, and Technical, with a Recommendatory Preface by Dr. Hofmann. In one large octavo volume of 662 pages, with illustrations.

---

ASHWELL (SAMUEL), M. D.—A PRACTICAL TREATISE ON THE DISEASES PECULIAR TO WOMEN. Illustrated by Cases derived from Hospital and Private Practice. Third American edition. In one octavo volume of 520 pages.

---

ARNOTT (NEILL), M. D.—ELEMENTS OF PHYSICS; or, Natural Philosophy, General and Medical. Written for universal use, in plain or non-technical language. A new edition, by Isaac Hays, M. D. Complete in one octavo volume, of 484 pages, with about two hundred illustrations.

---

BROWN (ISAAC BAKER), M. D.—ON SOME DISEASES OF WOMEN ADMITTING OF SURGICAL TREATMENT. With handsome illustrations. 1 volume, 8vo., extra cloth.

---

BENNETT (J. HUGHES), M. D.—THE PATHOLOGY AND TREATMENT OF PULMONARY TUBERCULOSIS, and on the Local Medication of Pharyngeal and Laryngeal Diseases, frequently mistaken for, or associated with, Phthisis. In one handsome octavo volume, with beautiful wood-cuts.

---

BENNETT (HENRY), M. D.—A PRACTICAL TREATISE ON INFLAMMATION OF THE UTERUS, ITS CERVIX AND APPENDAGES, and on its Connection with Uterine Disease. Fourth American, from the third and revised London edition. To which is added (*July*, 1856), A REVIEW OF THE PRESENT STATE OF UTERINE PATHOLOGY. In one neat octavo volume, of 500 pages, with wood-cuts.

---

BEALE (LIONEL JOHN), M. R. C. S.—THE LAWS OF HEALTH IN RELATION TO MIND AND BODY. A Series of Letters from an old Practitioner to a Patient. In one handsome volume, royal 12mo., extra cloth.

---

BILLING (ARCHIBALD) M. D.—THE PRINCIPLES OF MEDICINE. Second American, from the fifth and improved London edition. In one handsome octavo volume, extra cloth, 250 pp.

---

BARCLAY (A. W.), M. D.—A MANUAL OF MEDICAL DIAGNOSIS; being an Analysis of the Signs and Symptoms of Disease. In one neat octavo volume, extra cloth, of over 400 pages. (*Now Ready.*)

---

BUCKNILL (J. C.), M. D., and DANIEL H. TUKE, M. D.—A MANUAL OF PSYCHOLOGICAL MEDICINE; containing the History, Nosology, Description, Statistics, Diagnosis, Pathology, and Treatment of Insanity. With a Plate. In one handsome octavo volume, of 536 pages. (*Now Ready.*)

BLAKISTON (PEYTON), M. D.—PRACTICAL OBSERVATIONS ON CERTAIN DISEASES OF THE CHEST, and on the Principles of Auscultation. In one volume, 8vo., 384 pages.

BURROWS (GEORGE), M. D.—ON DISORDERS OF THE CEREBRAL CIRCULATION, and on the Connection between the Affections of the Brain and Diseases of the Heart. In one 8vo. vol., with colored plates, pp. 216.

BUDD (GEORGE), M. D.—ON DISEASES OF THE LIVER. Third American, from the third and enlarged London edition. In one very handsome octavo volume, with four beautifully colored plates, and numerous wood-cuts. 500 pages. New edition.

BUDD (GEORGE), M. D.—ON THE ORGANIC DISEASES AND FUNCTIONAL DISORDERS OF THE STOMACH. In one neat octavo volume, extra cloth. (*Now Ready.*)

BUCKLER (T. H.), M. D.—ON THE ETIOLOGY, PATHOLOGY, AND TREATMENT OF FIBRO-BRONCHITIS AND RHEUMATIC PNEUNOMIA. In one handsome octavo volume, extra cloth.

BUSHNAN (J. S.), M. D.—PRINCIPLES OF ANIMAL AND VEGETABLE PHYSIOLOGY. A Popular Treatise on the Functions and Phenomena of Organic Life. In one handsome royal 12mo. volume, extra cloth, with numerous illustrations.

BLOOD AND URINE (MANUALS ON).—BY JOHN WILLIAM GRIFFITH, G. OWEN REESE, AND ALFRED MARKWICK. One thick volume, royal 12mo., extra cloth, with plates. 460 pages.

BRODIE (SIR BENJAMIN C.), M. D.—CLINICAL LECTURES ON SURGERY. One vol., 8vo., cloth. 350 pages.

BIRD (GOLDING), M. D.—URINARY DEPOSITS: THEIR DIAGNOSIS, PATHOLOGY, AND THERAPEUTICAL INDICATIONS. A new and enlarged American, from the last improved London edition. With over sixty illustrations. In one royal 12mo. volume, extra cloth.

BARTLETT (ELISHA), M. D.—THE HISTORY, DIAGNOSIS, AND TREATMENT OF THE FEVERS OF THE UNITED STATES. Fourth edition, revised, with Additions by Alonzo Clark, M. D. In one handsome octavo volume, of 600 pages. (*Just Issued.*)

BOWMAN (JOHN E.), M. D.—PRACTICAL HANDBOOK OF MEDICAL CHEMISTRY. Second American, from the third and revised London edition. In one neat volume, royal 12mo., with numerous illustrations. 288 pages.

BOWMAN (JOHN E.), M. D.—INTRODUCTION TO PRACTICAL CHEMISTRY, INCLUDING ANALYSIS. Second American, from the second and revised English edition. With numerous illustrations. In one neat volume, royal 12mo. 350 pages.

BARLOW (GEORGE H.), M. D.—A MANUAL OF THE PRACTICE OF MEDICINE. With Additions by D. F. Condie, M. D. In one handsome octavo volume, leather, of 600 pages.

CURLING (T. B.), F. R. S.—A PRACTICAL TREATISE ON DISEASES OF THE TESTIS, SPERMATIC CORD, AND SCROTUM. Second American, from the second and enlarged English edition. With numerous illustrations. In one handsome octavo volume, extra cloth.

COLOMBAT DE L'ISERE.—A TREATISE ON THE DISEASES OF FEMALES, and on the Special Hygiene of their Sex. Translated, with many Notes and Additions, by C. D. Meigs, M. D. Second edition, revised and improved. In one large volume, octavo, with numerous woodcuts. 720 pages.

COPLAND (JAMES), M. D.—OF THE CAUSES, NATURE, AND TREATMENT OF PALSY AND APOPLEXY, and of the Forms, Seats, Complications, and Morbid Relations of Paralytic and Apoplectic Diseases. In one volume, royal 12mo., extra cloth. 326 pages

CARSON (JOSEPH), M. D.—SYNOPSIS OF THE COURSE OF LECTURES ON MATERIA MEDICA AND PHARMACY, delivered in the University of Pennsylvania. Second edition, revised. In one very neat octavo volume, of 208 pages.

---

CARPENTER (WILLIAM B.), M. D.—PRINCIPLES OF HUMAN PHYSIOLOGY; with their chief applications to Psychology, Pathology, Therapeutics, Hygiene, and Forensic Medicine. A new American, from the last and revised London edition. With nearly three hundred illustrations. Edited, with Additions, by Francis Gurney Smith, M. D., Professor of the Institutes of Medicine in the Pennsylvania Medical College, etc. In one very large and beautiful octavo volume, of about 900 large pages, handsomely printed, and strongly bound in leather, with raised bands. *(Just Issued.)*

---

CARPENTER (WILLIAM B.), M. D.—PRINCIPLES OF COMPARATIVE PHYSIOLOGY. New American, from the fourth and revised London edition. In one large and handsome octavo volume, with over three hundred beautiful illustrations.

---

CARPENTER (WILLIAM B.), M. D.—THE MICROSCOPE AND ITS REVELATIONS. With an Appendix containing the Applications of the Microscope to Clinical Medicine, by F. G. Smith, M. D. With 434 beautiful wood engravings. In one large and very handsome octavo volume of 724 pages, extra cloth or leather. *(Now Ready.)*

---

CARPENTER (WILLIAM B.), M. D.—ELEMENTS (OR MANUAL) OF PHYSIOLOGY, INCLUDING PHYSIOLOGICAL ANATOMY. Second American, from a new and revised London edition. With one hundred and ninety illustrations. In one very handsome octavo volume.

---

CARPENTER (WILLIAM B.), M. D.—PRINCIPLES OF GENERAL PHYSIOLOGY, INCLUDING ORGANIC CHEMISTRY AND HISTOLOGY. With a General Sketch of the Vegetable and Animal Kingdom. In one large and handsome octavo volume, with several hundred illustrations. *(Preparing.)*

---

CARPENTER (WILLIAM B.), M. D.—A PRIZE ESSAY ON THE USE OF ALCOHOLIC LIQUORS IN HEALTH AND DISEASE. New edition, with a Preface by D. F. Condie, M. D., and explanations of scientific words. In one neat 12mo. volume.

---

CHRISTISON (ROBERT), M. D.—A DISPENSATORY; or, Commentary on the Pharmacopœias of Great Britain and the United States: comprising the Natural History, Description, Chemistry, Pharmacy, Actions, Uses, and Doses of the Articles of the Materia Medica. Second edition, revised and improved, with a Supplement containing the most important New Remedies. With copious Additions, and two hundred and thirteen large wood-engravings. By R. Eglesfeld Griffith, M. D. In one very large and handsome octavo volume, of over 1000 pages.

---

CHELIUS (J. M.), M. D.—A SYSTEM OF SURGERY. Translated from the German, and accompanied with additional Notes and References, by John F. South. Complete in three very large octavo volumes, of nearly 2200 pages, strongly bound, with raised bands and double titles.

---

CONDIE (D. F.), M. D.—A PRACTICAL TREATISE ON THE DISEASES OF CHILDREN. Fifth edition revised and augmented. In one large volume, 8vo., of over 750 pages. *(Now Ready.)*

---

COOPER (BRANSBY B.), M. D.—LECTURES ON THE PRINCIPLES AND PRACTICE OF SURGERY. In one very large octavo volume, of 750 pages.

---

COOPER (SIR ASTLEY P.)—A TREATISE ON DISLOCATIONS AND FRACTURES OF THE JOINTS. Edited by Bransby B. Cooper, F. R. S., etc. With additional Observations by Prof. J. C. Warren. A new American edition. In one octavo volume, with numerous wood-cuts.

COOPER (SIR ASTLEY P.)—On the Structure and Diseases of the Testis, and on the Thymus Gland. One vol. imperial 8vo., with 177 figures, on 29 plates.

COOPER (SIR ASTLEY P.)—On the Anatomy and Diseases of the Breast, with twenty-five Miscellaneous and Surgical Papers. One large volume, imperial 8vo., with 252 figures, on 36 plates

CHURCHILL (FLEETWOOD), M. D.—On the Theory and Practice of Midwifery. A new American, from the last and improved English edition. Edited, with Notes and Additions, by D. Francis Condie, M. D., author of a "Practical Treatise on the Diseases of Children," etc. With 139 illustrations. In one very handsome octavo volume, 510 pages.

CHURCHILL (FLEETWOOD), M. D.—On the Diseases of Infants and Children. Second American edition, revised and enlarged by the author. With Additions by W. V. Keating, M. D. In one large and handsome volume of 700 pages. (*Now Ready.*)

CHURCHILL (FLEETWOOD), M. D.—Essays on the Puerperal Fever, and other Diseases peculiar to Women. Selected from the writings of British authors previous to the close of the eighteenth century. In one neat octavo volume, of about 450 pages.

CHURCHILL (FLEETWOOD), M. D.—On the Diseases of Women; including those of Pregnancy and Childbed. A new American edition, revised by the author. With Notes and Additions, by D. Francis Condie, M. D., author of a "Practical Treatise on the Diseases of Children." In one handsome octavo volume, of 768 pages, with wood-cuts. (*Now Ready.*)

DEWEES (W. P.), M. D.—A Comprehensive System of Midwifery. Illustrated by occasional Cases and many Engravings. Twelfth edition, with the Author's last Improvements and Corrections. In one octavo volume, of 600 pages.

DEWEES (W. P.), M. D.—A Treatise on the Physical and Medical Treatment of Children. Tenth edition. In one volume, octavo, 548 pages.

DEWEES (W. P.), M. D.—A Treatise on the Diseases of Females. Tenth edition. In one volume, octavo, 532 pages, with plates.

DRUITT (ROBERT), M. R. C. S.—The Principles and Practice of Modern Surgery. A new American, from the improved London edition. Edited by F. W. Sargent, M. D., author of "Minor Surgery," &c. Illustrated with one hundred and ninety-three wood-engravings In one very handsomely-printed octavo volume, of 576 large pages.

DUNGLISON, FORBES, TWEEDIE, AND CONOLLY.—The Cyclopædia of Practical Medicine: comprising Treatises on the Nature and Treatment of Diseases, Materia Medica and Therapeutics, Diseases of Women and Children, Medical Jurisprudence, &c. &c. In four large super-royal octavo volumes, of 3254 double-columned pages, strongly and handsomely bound.

*** This work contains no less than four hundred and eighteen distinct treatises, contributed by sixty-eight distinguished physicians.

DUNGLISON (ROBLEY), M. D.—Medical Lexicon; a Dictionary of Medical Science, containing a concise Explanation of the various Subjects and Terms of Physiology, Pathology, Hygiene, Therapeutics, Pharmacology, Obstetrics, Medical Jurisprudence, &c. With the French and other Synonymes; Notices of Climate and of celebrated Mineral Waters; Formulæ for various Officinal, Empirical, and Dietetic Preparations, &c. Fifteenth edition, revised. In one very thick octavo volume, of about 1000 large double-columned pages, strongly bound in leather, with raised bands. (*Now Ready.*)

DUNGLISON (ROBLEY), M. D.—THE PRACTICE OF MEDICINE. A Treatise on Special Pathology and Therapeutics. Third edition. In two large octavo volumes, of 1500 pages.

DUNGLISON (ROBLEY), M. D.—GENERAL THERAPEUTICS AND MATERIA MEDICA; adapted for a Medical Text-book. Sixth edition, much improved. With one hundred and eighty-seven Illustrations. In two large and handsomely printed octavo volumes, of about 1100 pages. (*Just Issued.*)

DUNGLISON (ROBLEY), M. D.—NEW REMEDIES, WITH FORMULÆ FOR THEIR PREPARATION AND ADMINISTRATION. Seventh Edition, with extensive Additions. In one very large octavo volume, of 770 pages. (*Now Ready.*)

DUNGLISON (ROBLEY), M. D.—HUMAN PHYSIOLOGY. Eighth edition. Thoroughly revised and extensively modified and enlarged, with over 500 illustrations. In two large and handsomely printed octavo volumes, containing about 1500 pages.

DICKSON (S. H.), M. D.—ELEMENTS OF MEDICINE: a Compendious View of Pathology and Therapeutics, or the History and Treatment of Diseases. In one large and handsome octavo volume of 750 pages, leather. (*Just Issued.*)

DAY (GEORGE E.), M. D.—A PRACTICAL TREATISE ON THE DOMESTIC MANAGEMENT AND MORE IMPORTANT DISEASES OF ADVANCED LIFE. With an Appendix on a new and successful mode of treating Lumbago and other forms of Chronic Rheumatism. One volume octavo, 226 pages.

ELLIS (BENJAMIN), M. D.—THE MEDICAL FORMULARY; being a Collection of Prescriptions, derived from the writings and practice of many of the most eminent physicians of America and Europe. Together with the usual Dietetic Preparations and Antidotes for Poisons. To which is added an Appendix on the Endermic use of Medicines, and on the use of Ether and Chloroform. The whole accompanied with a few brief Pharmaceutic and Medical Observations. Tenth edition, revised and much extended. by Robert P. Thomas, M.D., Professor of Materia Medica in the Philadelphia College of Pharmacy. In one neat octavo volume of 296 pages.

ERICHSEN (JOHN).—THE SCIENCE AND ART OF SURGERY; being a Treatise on Surgical Injuries, Diseases, and Operations. With Notes and Additions by the American editor. Illustrated with over 300 engravings on wood. In one large and handsome octavo volume of nearly 900 closely printed pages.

FLINT (AUSTIN), M. D.—PHYSICAL EXPLORATION AND DIAGNOSIS OF DISEASES AFFECTING THE RESPIRATORY ORGANS. In one handsome octavo volume, extra cloth, of 636 pages. (*Now Ready.*)

FERGUSSON (WILLIAM), F. R. S.—A SYSTEM OF PRACTICAL SURGERY. Fourth American, from the third and enlarged London edition. In one large and beautifully printed octavo volume of about 700 pages, with 393 handsome illustrations.

FRICK (CHARLES), M. D.—RENAL AFFECTIONS: their Diagnosis and Pathology. With illustrations. One volume, royal 12mo., extra cloth.

FOWNES (GEORGE), Ph. D. — ELEMENTARY CHEMISTRY, Theoretical and Practical. With numerous Illustrations. Edited, with Additions, by Robert Bridges, M. D. In one large royal 12mo. volume, of over 550 pages, with 181 wood-cuts. Sheep, or extra cloth.

FISKE FUND PRIZE ESSAYS ON CONSUMPTION.—By EDWIN LEE and EDWARD WARREN, M. D In one neat octavo volume, extra cloth. (*Now Ready.*)

GRAHAM (THOMAS), F. R. S.—THE ELEMENTS OF INORGANIC CHEMISTRY. Including the Application of the Science to the Arts. With numerous illustrations. With Notes and Additions, by Robert Bridges, M.D., etc., etc. Second American, from the second and enlarged London edition. 1 vol. 8vo., of over 800 pages, extra cloth. $4.00.

PART I. (*lately issued*), large 8vo., 430 pages, 185 illustrations. $1.50.
PART II. (*now ready*), to match, over 400 pages. $2.50.

---

GROSS (SAMUEL D.), M. D.—A PRACTICAL TREATISE ON THE DISEASES, INJURIES, AND MALFORMATIONS OF THE URINARY BLADDER, THE PROSTATE GLAND, AND THE URETHRA. Second edition, revised and much enlarged, with 184 illustrations. In one very large and handsome octavo volume, of over 900 pages, extra cloth or leather. (*Just Issued.*)

---

GROSS (SAMUEL D.), M.D.—A PRACTICAL TREATISE ON FOREIGN BODIES IN THE AIR-PASSAGES. In one handsome octavo volume, with illustrations.

---

GROSS (SAMUEL D.), M. D.—ELEMENTS OF PATHOLOGICAL ANATOMY; illustrated with three hundred and fifty wood-cuts. Third and revised edition. In one large octavo volume, of over 900 pages, leather.

---

GROSS (SAMUEL D.), M. D.—A SYSTEM OF SURGERY; Diagnostic, Pathological, Therapeutic, and Operative. With very numerous engravings on wood. (*Preparing.*)

---

GLUGE (GOTTLIEB), M. D.—AN ATLAS OF PATHOLOGICAL HISTOLOGY. Translated, with Notes and Additions, by Joseph Leidy, M. D., Professor of Anatomy in the University of Pennsylvania. In one volume, very large imperial quarto, with 320 figures, plain and colored, on twelve copper-plates.

---

GRIFFITH (ROBERT E.), M. D.—A UNIVERSAL FORMULARY, containing the methods of Preparing and Administering Officinal and other Medicines. The whole adapted to Physicians and Pharmaceutists. Second edition, thoroughly revised, with numerous Additions, by Robert P. Thomas, M. D., Professor of Materia Medica in the Philadelphia College of Pharmacy. In one large and handsome octavo volume of over 600 pages, double columns.

---

GRIFFITH (ROBERT E.), M. D.—MEDICAL BOTANY; or, a Description of all the more important Plants used in Medicine, and of their properties, Uses, and Modes of Administration. In one large octavo volume of 704 pages, handsomely printed, with nearly 350 illustrations on wood.

---

GARDNER (D. PEREIRA), M. D.—MEDICAL CHEMISTRY, for the use of Students and the Profession: being a Manual of the Science, with its Applications to Toxicology, Physiology, Therapeutics, Hygiene, &c. In one handsome royal 12mo. volume, with illustrations.

---

HARRISON (JOHN), M. D.—AN ESSAY TOWARDS A CORRECT THEORY OF THE NERVOUS SYSTEM. In one octavo volume, 292 pages.

---

HUGHES (H. M.), M. D.—A CLINICAL INTRODUCTION TO THE PRACTICE OF AUSCULTATION, and other Modes of Physical Diagnosis, in Diseases of the Lungs and Heart. Second American from the second and improved London edition. In one royal 12mo. volume.

---

HORNER (WILLIAM E.), M. D.—SPECIAL ANATOMY AND HISTOLOGY. Eighth edition. Extensively revised and modified. In two large octavo volumes, of more than 1000 pages, handsomely printed, with over 300 illustrations.

---

HOBLYN (RICHARD D.), A. M.—A DICTIONARY OF THE TERMS USED IN MEDICINE AND THE COLLATERAL SCIENCES. Second and improved American edition. Revised, with numerous Additions, from the second London edition, by Isaac Hays, M. D., &c. In one large royal 12mo. volume, of over 500 pages, double columns. (*Now Ready.*)

---

HABERSHON (S. O.), M. D.—PATHOLOGICAL AND PRACTICAL OBSERVATIONS ON DISEASES OF THE ALIMENTARY CANAL, ŒSOPHAGUS, STOMACH, CÆCUM, AND INTESTINES. With illustrations on wood. In one handsome octavo volume.

*** Publishing in the "Medical News and Library" for 1858.

HAMILTON (FRANK H.)—A TREATISE ON FRACTURES AND DISLOCATIONS. In one handsome octavo volume. With numerous illustrations. (*Preparing.*)

---

HOLLAND (SIR HENRY), M. D.—MEDICAL NOTES AND REFLECTIONS. From the third London edition. In one handsome octavo volume, extra cloth, of about 500 pages. (*Just Issued.*)

---

JONES (T. WHARTON), F. R. S.—THE PRINCIPLES AND PRACTICE OF OPHTHALMIC MEDICINE AND SURGERY. Second American, from the second and revised English edition. With Additions by Edward Hartshorne, M. D. In one very neat volume, large royal 12mo., of 500 pages, with 110 illustrations.

---

JONES (C. HANDFIELD), F. R. S., AND EDWARD H. SIEVEKING, M. D.—A MANUAL OF PATHOLOGICAL ANATOMY. With 397 engravings on wood. In one handsome volume, octavo, of nearly 750 pages, leather. (*Lately Issued.*)

---

KIRKES (WILLIAM SENHOUSE), M. D., AND JAMES PAGET, F. R. S.—A MANUAL OF PHYSIOLOGY. Third American, from the third and improved London edition. With 200 illustrations. In one large and handsome royal 12mo. volume. 586 pages. (*Now Ready.*)

---

KNAPP (F.), PH. D.—TECHNOLOGY; or, Chemistry applied to the Arts and to Manufactures. Edited, with numerous Notes and Additions, by Dr. Edmund Ronalds and Dr. Thomas Richardson. First American edition, with Notes and Additions, by Professor Walter R. Johnson. In two handsome octavo volumes, printed and illustrated in the highest style of art, with about 500 wood-engravings.

---

LEHMANN (G. C.).—PHYSIOLOGICAL CHEMISTRY. Translated from the second edition by George E. Day, M. D. Edited by R. E. Rogers, M. D. With illustrations selected from Funke's Atlas of Physiological Chemistry, and an Appendix of Plates. Complete in two handsome octavo volumes, extra cloth, containing 1200 pages. With nearly 200 illustrations. (*Just Issued.*)

---

LEHMANN (G. C.).—MANUAL OF CHEMICAL PHYSIOLOGY. Translated from the German, with Notes and Additions, by J. C. Morris, M. D. With an introductory Essay on Vital Force, by Samuel Jackson, M. D. In one handsome octavo volume, extra cloth, of 336 pages. With numerous illustrations.

---

LA ROCHE (R.), M. D.—PNEUMONIA; its Supposed Connection, Pathological and Etiological, with Autumnal Fevers, including an Inquiry into the Existence and Morbid Agency of Malaria. In one handsome octavo volume, extra cloth, of 500 pages.

---

LA ROCHE (R.), M. D.—YELLOW FEVER, considered in its Historical, Pathological, Etiological, and Therapeutical Relations. Including a Sketch of the Disease as it has occurred in Philadelphia from 1699 to 1854, with an Examination of the Connections between it and the Fevers known under the same name in other Parts of Temperate, as well as in Tropical Regions. In two large and handsome octavo volumes, of nearly 1500 pages, extra cloth. (*Just Issued.*)

---

LAWRENCE (W.), F. R. S.—A TREATISE ON DISEASES OF THE EYE. A new edition, edited, with numerous Additions, and 243 illustrations, by Isaac Hays, M. D., Surgeon to Wills' Hospital, etc. In one very large and handsome octavo volume of 950 pages, strongly bound in leather, with raised bands.

---

LUDLOW (J. L.), M. D.—A MANUAL OF EXAMINATIONS UPON ANATOMY, PHYSIOLOGY, SURGERY, PRACTICE OF MEDICINE, OBSTETRICS, MATERIA MEDICA, CHEMISTRY, PHARMACY, AND THERAPEUTICS. To which is added, a Medical Formulary. Third edition, thoroughly revised and greatly extended and enlarged. With 370 illustrations. In one large and handsome royal 12mo. volume, leather, of over 800 pages. (*Just Issued.*)

---

LAYCOCK (THOMAS), M. D.—LECTURES ON THE PRINCIPLES AND METHODS OF MEDICAL OBSERVATION AND RESEARCH. In one very neat royal 12mo. volume, extra cloth. (*Just Issued.*)

LALLEMAND (M.).—THE CAUSES, SYMPTOMS, AND TREATMENT OF SPERMATORRHŒA. Translated and edited by Henry J. McDougal. In one volume, octavo, of 320 pages. Second American edition.

LARDNER (DIONYSIUS), D. C. L.—HANDBOOKS OF NATURAL PHILOSOPHY AND ASTRONOMY. Revised, with numerous Additions, by the American editor. The whole complete in three volumes, 12mo., of about 2000 large pages, with over 1000 figures on steel and wood.

MORLAND (W. W.), M. D.—DISEASES OF THE URINARY ORGANS; a Compendium of their Diagnosis, Pathology, and Treatment. With Illustrations. In one large and handsome octavo volume, of about 600 pages, extra cloth. (*Now Ready*, Oct., 1858.)

MEIGS (CHARLES D.), M. D.—WOMAN: HER DISEASES AND THEIR REMEDIES. A Series of Lectures to his Class. Third and improved edition. In one large and beautifully-printed octavo volume.

MEIGS (CHARLES D.), M. D.—OBSTETRICS: THE SCIENCE AND THE ART. Second edition, revised and improved. With 131 illustrations. In one beautifully-printed octavo volume, of 752 large pages.

MEIGS (CHARLES D.), M. D.—A TREATISE ON ACUTE AND CHRONIC DISEASES OF THE NECK OF THE UTERUS. With numerous plates, drawn and colored from nature, in the highest style of art. In one handsome octavo volume, extra cloth.

MEIGS (CHARLES D.), M. D.—ON THE NATURE, SIGNS, AND TREATMENT OF CHILDBED FEVER, in a Series of Letters addressed to the Students of his Class. In one handsome octavo volume, extra cloth, of 365 pages.

MILLER (JAMES), F. R. S. E.—PRINCIPLES OF SURGERY. Fourth American, from the third and revised Edinburgh edition. In one large and very beautiful volume of 700 pages, with 240 exquisite illustrations on wood.

MILLER (JAMES), F. R. S. E.—THE PRACTICE OF SURGERY. Fourth American, from the third Edinburgh edition. Edited, with Additions. Illustrated by 364 engravings on wood. In one large octavo volume of over 700 pages.

MALGAIGNE (J. F.).—OPERATIVE SURGERY, based on Normal and Pathological Anatomy. Translated from the French, by Frederick Brittan, A. B., M. D. With numerous illustrations on wood. In one handsome octavo volume of nearly 600 pages.

MOHR (FRANCIS), PH. D., AND REDWOOD (THEOPHILUS).—PRACTICAL PHARMACY. Comprising the Arrangements, Apparatus, and Manipulations of the Pharmaceutical Shop and Laboratory. Edited, with extensive Additions, by Prof. William Procter, of the Philadelphia College of Pharmacy. In one handsomely-printed octavo volume, of 570 pages, with over 500 engravings on wood.

MACLISE (JOSEPH).—SURGICAL ANATOMY. Forming one volume, very large imperial quarto. With sixty-eight large and splendid Plates, drawn in the best style, and beautifully colored. Containing 190 Figures, many of them the size of life. Together with copious and explanatory letter-press. Strongly and handsomely bound in extra cloth, being one of the cheapest and best executed Surgical works as yet issued in this country.

Copies can be sent by mail, in five parts, done up in stout covers.

MILLER (HENRY), M. D.—PRINCIPLES AND PRACTICE OF OBSTETRICS; including the Treatment of Chronic Inflammation of the Cervix and Body of the Uterus, considered as a frequent Cause of Abortion. With about 100 engravings on wood. In one very handsome octavo volume of over 600 pages. (*Now Ready*.)

MONTGOMERY (W. F.), M. D., &c.—AN EXPOSITION OF THE SIGNS AND SYMPTOMS OF PREGNANCY. With some other Papers on Subjects connected with Midwifery. From the second and enlarged English edition. With two exquisite coloured plates and numerous woodcuts. In one very handsome octavo volume, extra cloth, of nearly 600 pages.

MAYNE (JOHN), M. D.—A DISPENSATORY AND THERAPEUTICAL REMEMBRANCER. Comprising the entire lists of Materia Medica, with every Practical Formula contained in the three British Pharmacopœias. In one 12mo. volume, extra cloth, of over 300 large pages.

MACKENZIE (W.), M. D.—A PRACTICAL TREATISE ON DISEASES AND INJURIES OF THE EYE. To which is prefixed an Anatomical Introduction, by T. Wharton Jones. From the fourth revised and enlarged London edition. With Notes and Additions by Addinell Hewson, M. D. In one very large and handsome octavo volume, with numerous wood-cuts and plates. 1028 pages, leather, raised bands. (*Just Issued.*)

NEILL (JOHN), M. D., AND FRANCIS GURNEY SMITH, M. D.—AN ANALYTICAL COMPENDIUM OF THE VARIOUS BRANCHES OF MEDICAL SCIENCE; for the Use and Examination of Students. Second edition, revised and improved. In one very large and handsomely printed royal 12mo. volume of over 1000 pages, with 350 illustrations on wood. Strongly bound in leather, with raised bands.

NEILL (JOHN), M. D.—OUTLINES OF THE NERVES. 1 vol. 8vo., with handsome plates. OUTLINES OF THE VEINS AND LYMPHATICS, 1 vol. 8vo., handsome colored plates.

NELIGAN (J. MOORE), M. D.—ATLAS OF CUTANEOUS DISEASES. In one beautiful quarto volume, extra cloth, with splendid colored plates, presenting nearly one hundred elaborate representations of disease. (*Now Ready.*)

NELIGAN (J. MOORE), M. D.—A PRACTICAL TREATISE ON DISEASES OF THE SKIN. In one neat royal 12mo. volume, of 334 pages.

OWEN (PROF. R.)—ON THE DIFFERENT FORMS OF THE SKELETON. One royal 12mo. volume, with numerous illustrations.

PARKER (LANGSTON).—THE MODERN TREATMENT OF SYPHILITIC DISEASES, BOTH PRIMARY AND SECONDARY; comprising the Treatment of Constitutional and Confirmed Syphilis, by a safe and successful method. With numerous Cases, Formulæ, and Clinical Observations. From the third and entirely rewritten London edition. In one neat octavo volume.

PEREIRA (JONATHAN), M. D.—THE ELEMENTS OF MATERIA MEDICA AND THERAPEUTICS. Third American edition, enlarged and improved by the author; including Notices of most of the Medical Substances in use in the civilized world, and forming an Encyclopædia of Materia Medica. Edited, with Additions, by Joseph Carson, M. D., Professor of Materia Medica and Pharmacy in the University of Pennsylvania. In two very large octavo volumes of 2100 pages, on small type, with over 450 illustrations. (*Now Complete.*)

PARRISH (EDWARD).—AN INTRODUCTION TO PRACTICAL PHARMACY. Designed as a Text-book for the Student, and as a Guide for the Physician and Pharmaceutist. With many Formulæ and Prescriptions. In one handsome octavo volume, extra cloth, of 550 pages, with 243 illustrations.

PEASELEE (E. R.), M. D.—HUMAN HISTOLOGY, in its Applications to Physiology and General Pathology. With 434 illustrations. In one handsome octavo volume. (*Now Ready.*)

PIRRIE (WILLIAM), F.R.S.E.—THE PRINCIPLES AND PRACTICE OF SURGERY. Edited by John Neill, M.D., Demonstrator of Anatomy in the University of Pennsylvania, Surgeon to the Pennsylvania Hospital, &c. In one very handsome octavo volume of 780 pages, with 316 illustrations.

RAMSBOTHAM (FRANCIS H.), M. D.—THE PRINCIPLES AND PRACTICE OF OBSTETRIC MEDICINE AND SURGERY, in reference to the Process of Parturition. A new and enlarged edition, thoroughly revised by the author. With Additions by W. V. Keating, M. D. In one large and handsome imperial octavo volume of 650 pages, strongly bound in leather, with raised bands. With sixty-four beautiful plates, and numerous wood-cuts in the text, containing in all nearly 200 large and beautiful figures. (*Just Issued.*)

RIGBY (EDWARD), M. D.—ON THE CONSTITUTIONAL TREATMENT OF FEMALE DISEASES. In one neat royal 12mo. volume, extra cloth, of about 250 pages. (*Just Issued.*)

RICORD (P.), M.D.—ILLUSTRATIONS OF SYPHILITIC DISEASE. Translated from the French, by Thomas F. Betton, M.D. With the addition of a History of Syphilis, and a complete Bibliography and Formulary of Remedies, collated and arranged by Paul B. Goddard, M.D. With fifty large quarto plates, comprising 117 beautifully colored illustrations. In one large and handsome quarto volume.

RICORD (P.), M.D.—A TREATISE ON THE VENEREAL DISEASE. By John Hunter, F.R.S. With copious Additions, by Ph. Ricord, M.D. Edited, with Notes, by Freeman J. Bumstead, M.D. In one handsome octavo volume, with plates.

RICORD (P.), M.D.—LETTERS ON SYPHILIS, addressed to the Chief Editor of the Union Médicale. With an Introduction, by Amédée Latour. Translated by W. P. Lattimore, M.D. In one neat octavo volume.

ROKITANSKY (CARL).—A MANUAL OF PATHOLOGICAL ANATOMY. Translated from the German by W. E. Swaine, Edward Sieveking, M.D., C. H. Moore, and George E. Day, M.D. Complete, four volumes bound in two, extra cloth, of about 1200 pages. (*Just Issued.*)

RIGBY (EDWARD), M.D.—A SYSTEM OF MIDWIFERY. With Notes and Additional Illustrations. Second American edition. One volume octavo, 422 pages.

ROYLE (J. FORBES), M.D.—MATERIA MEDICA AND THERAPEUTICS; including the Preparations of the Pharmacopœias of London, Edinburgh, Dublin, and of the United States. With many new Medicines. Edited by Joseph Carson, M.D., Professor of Materia Medica and Pharmacy in the University of Pennsylvania. With ninety-eight illustrations. In one large octavo volume of about 700 pages.

SKEY (FREDERICK C.), F.R.S.—OPERATIVE SURGERY. In one very handsome octavo volume of over 650 pages, with about 100 wood-cuts.

SHARPEY (WILLIAM), M.D., JONES QUAIN, M.D., AND RICHARD QUAIN, F.R.S., etc.—HUMAN ANATOMY. Revised, with Notes and Additions, by Joseph Leidy, M.D. Complete in two large octavo volumes, of about 1300 pages. Beautifully illustrated with over 500 engravings on wood.

SMITH (HENRY H.), M.D., AND WILLIAM E. HORNER, M.D.—AN ANATOMICAL ATLAS illustrative of the Structure of the Human Body. In one volume, large imperial octavo, with about 650 beautiful figures.

SMITH (HENRY H.), M.D.—MINOR SURGERY; or, Hints on the Every-day Duties of the Surgeon. With 247 illustrations. Third and enlarged edition. In one handsome royal 12mo. volume of 456 pages

SARGENT (F. W.), M.D.—ON BANDAGING AND OTHER OPERATIONS OF MINOR SURGERY. Second edition, enlarged. In one handsome royal 12mo. volume of nearly 400 pages, with 182 illustrations. (*Just Issued.*)

TILLÉ (ALFRED), M.D.—PRINCIPLES OF THERAPEUTICS. In one handsome volume. (*Preparing.*)

SIMON (JOHN), F.R.S.—GENERAL PATHOLOGY, as conducive to the Establishment of Rational Principles for the Prevention and Cure of Disease. A Course of Lectures delivered at St. Thomas's Hospital during the Summer Session of 1850. In one neat octavo volume.

SMITH (W. TYLER), M.D.—ON PARTURITION, AND THE PRINCIPLES AND PRACTICE OF OBSTETRICS. In one large duodecimo volume of 400 pages.

SMITH (W. TYLER), M.D.—THE PATHOLOGY AND TREATMENT OF LEUCORRHŒA. With numerous illustrations. In one very handsome octavo volume, extra cloth, of about 250 pages.

SOLLY (SAMUEL), F. R. S. — THE HUMAN BRAIN; its Structure, Physiology, and Diseases With a Description of the Typical Forms of the Brain in the Animal Kingdom. From the Second and much enlarged London edition. In one octavo volume, with 120 wood-cuts.

SCHŒDLER (FRIEDRICH), PH. D.—THE BOOK OF NATURE; an Elementary Introduction to the Sciences of Physics, Astronomy, Chemistry, Mineralogy, Geology, Botany, Zoology, and Physiology. First American edition, with a Glossary and other Additions and Improvements; from the second English edition. Translated from the sixth German edition, by Henry Medlock, F.C.S., &c. In one thick volume, small octavo, of about 700 pages, with 679 illustrations on wood. Suitable for the higher schools and private students. (*Now Ready.*)

TAYLOR (ALFRED S.), M. D., F. R. S.—MEDICAL JURISPRUDENCE. Fourth American, from the fifth and improved English edition. With Notes and References to American Decisions, by Edward Hartshorne, M. D. In one large octavo volume of 700 pages. (*Now Ready.*)

TAYLOR (ALFRED S.), M. D.—ON POISONS, IN RELATION TO MEDICAL JURISPRUDENCE AND MEDICINE. Edited, with Notes and Additions, by R. E. Griffith, M. D. In one large octavo volume of 688 pages.

TANNER (T. H.), M. D.—A MANUAL OF CLINICAL MEDICINE AND PHYSICAL DIAGNOSIS. To which is added, The Code of Ethics of the American Medical Association. In one neat volume, small 12mo., extra cloth, or flexible. (*Just Issued.*)

TODD (R. B.), M. D.—CLINICAL LECTURES ON CERTAIN DISEASES OF THE URINARY ORGANS, AND ON DROPSIES. In one octavo volume, extra cloth, of about 300 pages. (*Just Issued.*)

TODD (R. B.), M. D., AND WILLIAM BOWMAN, F. R. S.—PHYSIOLOGICAL ANATOMY AND PHYSIOLOGY OF MAN. Now complete, in one very large and handsome octavo volume, of 926 pages, with 300 illustrations on wood. (*Just Issued*, 1857.)

☞ Gentlemen who have the earlier portions of this work can still complete their copies, if early application be made.

WATSON (THOMAS), M. D., &c. — LECTURES ON THE PRINCIPLES AND PRACTICE OF PHYSIC. A new American, from the last London edition. Revised, with Additions, by D. Francis Condie, M. D., author of a "Treatise on the Diseases of Children," &c. In one imperial octavo volume, of over 1200 large pages, with nearly 200 cuts, strongly bound, with raised bands. (*Now Ready.*)

WALSHE (W. H.), M. D.—DISEASES OF THE HEART, LUNGS, AND APPENDAGES; their Symptoms and Treatment. In one handsome volume, large royal 12mo., 512 pages.

WHAT TO OBSERVE AT THE BEDSIDE AND AFTER DEATH, IN MEDICAL CASES. Published under the authority of the London Society for Medical Observation. In one very handsome volume, royal 12mo., extra cloth.

WILDE (W. R.).—AURAL SURGERY, AND THE NATURE AND TREATMENT OF DISEASES OF THE EAR. In one handsome octavo volume, with illustrations.

WHITEHEAD (JAMES), F. R. C. S., &c. — THE CAUSES AND TREATMENT OF ABORTION AND STERILITY; being the Result of an Extended Practical Inquiry into the Physiological and Morbid Conditions of the Uterus. Second American Edition. In one volume, octavo, 368 pages

WEST (CHARLES), M. D. — LECTURES ON THE DISEASES OF INFANCY AND CHILDHOOD. Second American, from the second and enlarged London edition. In one volume, octavo, of nearly 500 pages.

WEST (CHARLES), M. D.—AN INQUIRY INTO THE PATHOLOGICAL IMPORTANCE OF ULCERATION OF THE OS UTERI. Being the Croonian Lectures for the year 1854. In one neat octavo volume, extra cloth.

WEST (CHARLES), M. D. — LECTURES ON THE DISEASES OF WOMEN. In two Parts. Part I., Diseases of the Uterus. Part II., Diseases of the Ovaries, &c., the Bladder, Vagina, and External Organs. Complete in one octavo volume of 500 pages, extra cloth. (*Now Ready.*) Part II. now ready, 1 vol., 8vo., extra cloth, of about 200 pages. Sold separate, $1.

WILSON (ERASMUS), M. D., F. R. S.—A SYSTEM OF HUMAN ANATOMY, General and Special. Fourth American, from the last English edition. Edited by W. H. Gobrecht, M. D. With 400 illustrations. Beautifully printed, in one large octavo volume, of over 600 pages. (*Now Ready.*)

WILSON (ERASMUS), M. D., F. R. S.—THE DISSECTOR'S MANUAL; Practical and Surgical Anatomy. Third American, from the last revised and enlarged English edition. Modified and re-arranged by William Hunt, M. D. In one large and handsome royal 12mo. volume, leather, of 582 pages, with 154 illustrations.

WILSON (ERASMUS), M. D., F. R. S.—ON DISEASES OF THE SKIN. Fourth American, from the Fourth London edition. In one neat octavo volume, of 650 pages, extra cloth.

Also, AN ATLAS OF PLATES, of which twelve are exquisitely coloured, illustrating "WILSON ON THE SKIN." 8vo., cloth. (*Now Ready.*)

WILSON (ERASMUS), M. D., F. R. S.—ON CONSTITUTIONAL AND HEREDITARY SYPHILUS, AND ON SYPHILITIC ERUPTIONS. In one small octavo volume, beautifully printed, with four exquisite coloured plates, presenting more than thirty varieties of Syphilitic Eruptions.

WILSON (ERASMUS), M. D., F. R. S.—HEALTHY SKIN; a Treatise on the Management of the Skin and Hair in Relation to Health. Second American, from the fourth and improved London edition. In one handsome royal 12mo. volume, extra cloth, with numerous illustrations. Copies may also be had in paper covers, for mailing, price 75 cents.

WILLIAMS (C. J. B.), M.D., F.R.S.—PRINCIPLES OF MEDICINE; comprising General Pathology and Therapeutics, and a brief General View of Etiology, Nosology, Semeiology, Diagnosis, Prognosis, and Hygienics. Fifth American, from a new and enlarged London edition. In one octavo volume, of 500 pages.

YOUATT (WILLIAM), V. S. — THE HORSE. A new edition, with numerous illustrations; together with a General History of the Horse; a Dissertation on the American Trotting Horse; how Trained and Jockeyed; an account of his Remarkable Performances; and an Essay on the Ass and the Mule. By J. S. Skinner, formerly Assistant Postmaster-General, and Editor of the Turf Register. One large octavo volume.

YOUATT (WILLIAM), V. S.—THE DOG. Edited by E. J. Lewis, M. D. With numerous and beautiful illustrations. In one very handsome volume, crown 8vo., crimson cloth, gilt.

## Illustrated Catalogue.

Blanchard & Lea have now ready a detailed Catalogue of their publications, in Medical and other Sciences, with Specimens of the Wood-engravings, Notices of the Press, &c. &c., forming a pamphlet of eighty large octavo pages. It has been prepared without regard to expense, and may be considered as one of the handsomest specimens of printing as yet executed in this country. Copies will be sent free, by post, on receipt of nine cents in postage stamps.

Detailed Catalogues of their publications, Miscellaneous, Educational, Medical, &c., furnished gratis, on application.

www.ingramcontent.com/pod-product-compliance
Lightning Source LLC
LaVergne TN
LVHW020228110826
845151LV00003B/854
* 9 7 8 1 4 2 5 5 3 1 5 9 1 *